The Minerals, Metals & Materials Series

Yuri Hovanski · Yutaka Sato · Piyush Upadhyay ·
Anton A. Naumov · Nilesh Kumar
Editors

Friction Stir Welding and Processing XI

TMS

Springer

Editors
Yuri Hovanski
Brigham Young University
Provo, UT, USA

Yutaka Sato
Tohoku University
Sendai, Japan

Piyush Upadhyay
Pacific Northwest National Laboratory
Richland, WA, USA

Anton A. Naumov
Peter the Great St. Petersburg Polytechnic University
Saint Petersburg, Russia

Nilesh Kumar
The University of Alabama
Tuscaloosa, AL, USA

ISSN 2367-1181 ISSN 2367-1696 (electronic)
The Minerals, Metals & Materials Series
ISBN 978-3-030-65264-7 ISBN 978-3-030-65265-4 (eBook)
https://doi.org/10.1007/978-3-030-65265-4

Cover Illustration: Top left: From Chapter "Preliminary Investigation of the Effect of Temperature Control in Friction Stir Welding", Johnathon B. Hunt et al., Figure 2: Is an example of the weld setup for all temperature control welds. https://doi.org/10.1007/978-3-030-65265-4_8. Top right: From Chapter "Process Robustness of Friction Stir Dovetailing of AA7099 to Steel with In Situ AA6061 Interlayer Linking", Md Reza-E-Rabby et al., Figure 2: **a** Lap shear tensile strength as a function of applied forge force during FSD process, **b** example of FSD joint produced solely by mechanical interlocking, and **c** failed FSD tools (tips of WC inserts broken) at process force of 75 kN or more. https://doi.org/10.1007/978-3-030-65265-4_14. Bottom: From Chapter "Characterization and Analysis of Effective Wear Mechanisms on FSW Tools", Michael Hasieber et al., Figure 8: SEM analysis of the FSW tool in the initial state at the probe (**a**) and the shoulder (**b**) https://doi.org/10.1007/978-3-030-65265-4_3.

This Springer imprint is published by the registered company Springer Nature Switzerland AG
The registered company address is: Gewerbestrasse 11, 6330 Cham, Switzerland

Preface

These proceedings represent the 11th symposium on Friction Stir Welding and Processing (FSW/P) held under the auspices of The Minerals, Metals & Materials Society. These historic proceedings represent 30 years of study, research, and implementation since the initial FSW patent was filed in 1991. The continued interest and participation in this symposium and the associated proceedings are an indirect testimony of the growth of this field.

For 2021, a total of 61 abstracts were submitted, which include 8 oral sessions. There are 20 papers included in this volume, which when combined with the previous 10 proceedings' publications represent more than 325 papers over a 22-year period. These submissions cover all aspects of friction stir technologies including FSW of high melting temperature materials, FSW of lightweight materials, FSW of dissimilar materials, simulation of FSW/P, controls and inspection of FSW/P, and derivative technologies like friction stir processing, friction stir spot welding, additive friction stir, and friction stir extrusion.

Friction stir welding was invented by TWI (formerly The Welding Institute), Cambridge, UK and patented in 1991, although the real growth in this field started several years later. In the last 30 years, FSW has seen significant growth in both technology implementation and scientific exploration. The original patent has led to hundreds of additional patents issued globally, as various solid-state processing techniques have derived from the original FSW concept. In addition to the tremendous number of derivative technologies that have been developed based on the concept of friction stirring, thousands of papers have been published characterizing and documenting the commercial and scientific benefits of the same.

The organizers would like to thank the Shaping and Forming Committee of the Materials Processing and Manufacturing Division for sponsoring this symposium.

Yuri Hovanski
Yutaka Sato
Piyush Upadhyay
Anton A. Naumov
Nilesh Kumar

Contents

About the Editors

Yuri Hovanski is an Associate Professor of Manufacturing Engineering at Brigham Young University. He earned a B.S. degree in Mechanical Engineering at Brigham Young University, and then completed his masters and doctorate degrees at Washington State University. As a member of numerous professional societies, he actively participates in AWS, ASM, SME, and TMS serving in numerous leadership roles at the technical committee and division levels. He is the past chair of the TMS Shaping and Forming Committee, the chair of the ASM Joining of Advanced Specialty Materials Committee, and center director for the Center of Friction Stir Process, a retired NSF-IUCRC. He participated in research related to friction stir technologies for more than a decade as a senior research engineer at Pacific Northwest National Laboratory where he developed low-cost solutions for industrial implementation of friction stir technologies. Working with numerous industrial suppliers around the world, Dr. Hovanski has introduced cost-efficient solutions for thermal telemetry, demonstrated novel, low-cost tool materials for friction stir spot, implemented high volume production techniques for aluminum tailor-welded blanks, developed new methodologies for joining dissimilar materials, and reduced the cycle time for refill friction stir spot welding. As an active researcher, Dr. Hovanski received the R&D 100 award in 2011 and again in 2017, the DOE Vehicle Technologies Office Distinguished Achievement award in 2015, and a western region FLC award for technology transfer in 2015. He actively reviews friction stir related literature for numerous publications and has documented his work in more than 75 publications and proceedings and five U.S. patents.

Yutaka Sato is a Professor in the Department of Materials Processing at Tohoku University, Japan. He earned a Ph.D. in Materials Processing at Tohoku University (2001). His Ph.D. thesis was titled "Microstructural Study on Friction Stir Welds of Aluminum Alloys." He participated in friction stir research of steels at Brigham Young University for a year in 2003. He was a member of Sub-commission III-B WG-B4 at IIW, which is a working group to build international standardization of friction stir spot welding. His work has focused on metallurgical studies of friction stir welding and processing for more than 20 years. He has obtained fundamental knowledge on development of grain structure, texture evolution, joining mechanism, behavior of oxide-layer on surface, and properties–microstructure relationship. He has received a number of awards including the Kihara Award from the Association for Weld Joining Technology Promotion (2008), Prof. Koichi Masubuchi Award from AWS (2009), Murakami Young Researcher Award from the Japan Institute of Metals (2010), Honda Memorial Young Researcher Award (2011), The Japan Institute of Metals and Materials Meritorious Award (2015), and Light Metal Breakthrough Award from the Japan Institute of Light Metals (2017). He has authored or co-authored more than 250 papers in peer-reviewed journals and proceedings.

Piyush Upadhyay is a material scientist at Pacific Northwest National Laboratory. He earned his Ph.D. from the University of South Carolina in 2012 in "Boundary Condition Effects on Friction Stir Welding of Aluminum Alloys." For more than a decade he has been involved in research and development of FSW and allied technologies to join similar and dissimilar materials. Currently, he leads and contributes to projects on friction stir welding and joining of alloys in dissimilar thickness and dissimilar materials including combinations of aluminum, magnesium, steels, and polymers. He has received several awards and recognitions including Aid Nepal Scholarship for undergraduate study (2001),

Happy House Foundation Research Fellowship at Kathmandu University (2007), the DOE Energy Efficiency & Renewable Energy Recognition for Innovation (2015), and the R&D 100 award (2018). He has authored or coauthored more than 35 papers in peer-reviewed journals and proceedings and is actively involved as a guest editor and reviewer for technical journals and conference committees.

Anton A. Naumov is an Associate Professor at the Institute of Mechanical Engineering, Materials and Transport at Peter the Great St. Petersburg Polytechnic University (SPbPU). He obtained B.S. (2003), masters (2005), and doctorate (2010) degrees in Metallurgy at SPbPU. In 2012, 2014, and 2016 he won the Grants of the President of Russian Federation for young Ph.D.s in technical field of science. The 2016 grant was focused on the microstructure and mechanical properties evolution of aluminum alloys during friction stir welding. Dr. Naumov was one of the organizers for the Laboratory of Lightweight Materials and Structures (LWMS) at SPbPU in 2014 in terms of Mega-Grant of Ministry of Science and Education of Russian Federation for attracting leading foreign scientists and opening R&D labs under their supervision. He has been working in R&D laboratory LWMS until the present time as a principal researcher in the field of friction stir welding and processing. Dr. Naumov is an active member of the professional societies TMS and AWS and reviews friction stir related articles for several journals. He has authored or coauthored more than 40 papers in peer-reviewed journals and proceedings and has three Russian Federation patents.

Nilesh Kumar has worked as an Assistant Professor in the Department of Metallurgical and Materials Engineering at The University of Alabama since August 2018. Prior to this, he was a post-doctoral research associate in the Department of Nuclear Engineering at North Carolina State University Raleigh and the University of North Texas Denton. He obtained his Ph.D. degree in Materials Science and Engineering from Missouri University of Science & Technology Rolla. A common theme which cuts across all the research work Dr. Kumar has carried out so far is establishing correlation among processing (thermo-mechanical, friction stir processing, and laser), microstructure (grain-size, dislocations, and precipitates), and mechanical properties (strength, ductility, creep, creep-fatigue, residual stress, and stress corrosion cracking) of metallic materials (aluminum alloys, magnesium alloys, austenitic stainless steels, titanium alloys, high-entropy alloys, advanced high strength steel) used primarily in transportation and power-generation industries. He has published 38 papers in peer-reviewed journals, co-authored three books, and contributed two handbook chapters. Dr. Kumar is a member of several scientific organizations and has been a reviewer for more than 20 scientific journals. He is also the recipient of the Kent D. Peaslee Junior Faculty Award by the Association for Iron & Steel Technology (AIST) Foundation for the year 2019–2020.

Part I
Lightweight Materials and High Entropy Alloys

Case Study: Implementation of FSW in the Colombian Rail Transport Sector

Elizabeth Hoyos, Santiago Escobar, Jeroen De Backer, Jonathan Martin, and Mauricio Palacio

Abstract Medellín is the second-largest city in Colombia and the only one to have a metro system: Metro de Medellín (MdM), a mass-transport system that on average transports 811 thousand people daily. An exercise was carried out to identify components suitable for FSW and a *Tie*, part of the doors opening and closing mechanism, was selected. This component was originally manufactured from a bespoke extruded U-shaped AA6063-T83 profile with arc-welded features and machined details. FSW provides a solution for joining aluminum alloys compared to arc welding, but high investment costs for dedicated equipment remains an obstacle for its implementation in small workshops. To take advantage of FSW for small batches, a manufacturing-strategy was proposed based on the capabilities and limitations in the local metal-working sector, enabling implementation on adapted conventional milling machines. A representative demonstrator has been successfully manufactured for further testing. Finally, a cost comparison between the existing manufacturing route and FSW is presented.

Keywords Aluminum · Railway · Friction stir welding · Industrial application

E. Hoyos (✉) · S. Escobar
Mechanical Engineering Department, Universidad EIA, Envigado, Colombia
e-mail: elizabeth.hoyos@eia.edu.co

S. Escobar
e-mail: santiago.escobar23@eia.edu.co

J. De Backer · J. Martin
Friction & Forge Processes, TWI, Cambridge, UK
e-mail: jeroen.debacker@twi.co.uk

J. Martin
e-mail: jonathan.martin@twi.co.uk

M. Palacio
Research, Development and Innovation, Metro de Medellín, Medellín, Colombia
e-mail: mpalacio@metrodemedellin.gov.co

Y. Hovanski et al. (eds.), *Friction Stir Welding and Processing XI*, The Minerals, Metals & Materials Series, https://doi.org/10.1007/978-3-030-65265-4_1

Introduction

Since the late 2000s in Colombia, different authors have been working in the implementation of Friction stir welding using adapted machinery and tooling locally developed [1], aiming for the evaluation and selection of adequate parameters for 2XXX, 5XXX, and 7XXX aluminum alloys series [2–5], targeting the successful fabrication of sound welds. In recent years, new approaches around the technique have been made in mathematical modeling of the variables, assessing the residual stresses and development of new tools such as stationary shoulder FSW, trying to enhance the quality and performance with the available machinery [2–6]. Despite the knowledge, the implementation of FSW has not been successful for industrial components because there is an unfamiliarity by local engineers around the advantages and all the requirements of the procedure.

After decades of inactivity in Colombia, the railway industry flourished again [7], leading to new requirements of maintenance and fabrication of components. However, while there is a broad scientific knowledge of FSW, and a wide use in the railway industry [8–11], the execution of FSW by small workshops remains limited, and to date no FSW applications exist in Colombia. A broader knowledge around the considerations, designs, and mechanical analysis around welds is needed to exemplify the necessities and local limitations for particular applications. Due to the possibility of implementation of the technique in the railway industry and future growth projected, as means to educate the local engineers, a specific component of the aperture mechanism was evaluated of the Metro de Medellín (MdM); where considerations regarding its mechanical solicitations, machinery availability and design particularities were studied and implemented satisfactorily to build a prototype in accordance with the geometry requirements.

Materials and Methods

The component mentioned before is an 800 mm long "C"-shaped extrusion called *Tie*, with two distinctive geometrical features known as the *Bracket* and the *Lid*, both of them are shown in Fig. 1. Originally, this piece was manufactured as a tailored-made extruded profile of 80 × 60 × 5.5 mm but due to the age of the components, there is a lack of availability by the original manufacturer of the trains, finally affecting the operation of MdM.

There have been attempts to manufacture the *Tie* using a combination of existing techniques like machining and arc welding (over the dotted lines in Fig. 1), but the quality obtained does not fulfill the standards required. The discontinuities which jeopardize the performance of the pieces are related to cracks around the heat-affected zones, near the *Bracket* and *Lid*, combined with an improper surface finish; this can be observed in Fig. 2.

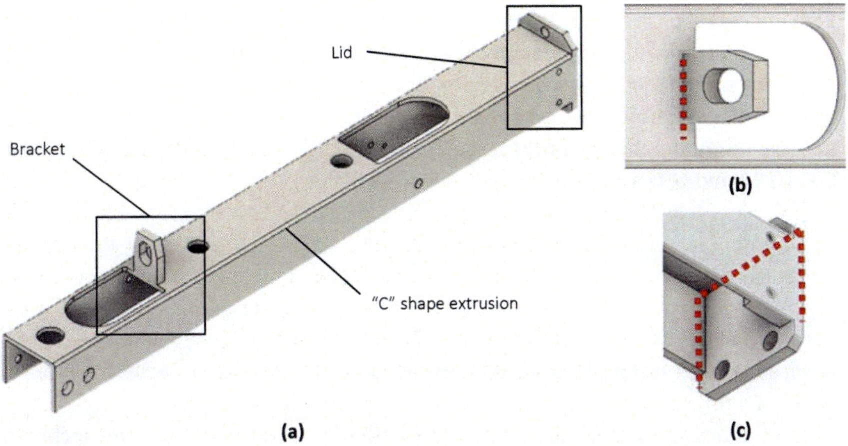

Fig. 1 Original component, **a** Tie, **b** Bracket zoom and **c** Lid zoom. (Color figure online)

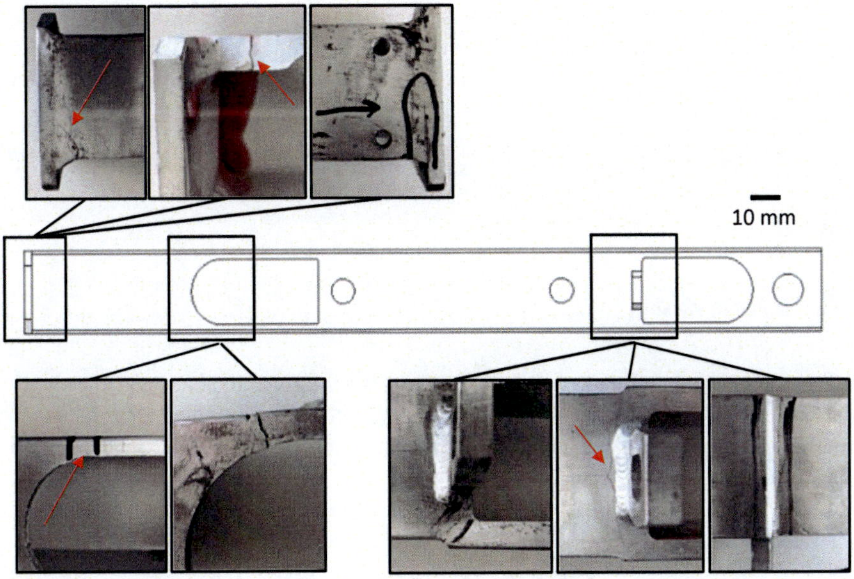

Fig. 2 Discontinuities observed in the components. Scale in millimeters. (Color figure online)

Two aluminium alloys were considered with similar or greater yield and ultimate strength, than the original alloys (214 MPa and 241 MPa, respectively [12]). These were AA6082-T6 (250 MPa/290 MPa [13]) and AA7075-T6 (503 MPa/572 MPa [14]).] The base metal selected for this application was the AA6082-T6 as it is widely available, very suitable for FSW commonly used in the railway applications, and has better corrosion resistance. For the proposed manufacturing strategy using

FSW, the U—profile was created from two extruded “L”-shaped profiles, which gives a solution with less material loss than machining from solid, and avoids the great investments in bespoke extrusion dies; which are not feasible considering the number of *Ties* projected to be manufactured are approximately 500, much lower than the minimum material needed to be extruded by a tailor-made die. Welded using 600 RPM and 600 mm/min

To assess feasibility of the FSW’ed component, the first step was a load analysis of the component to identify the maximum values of normal and shear stresses using a Von Mises analysis versus mechanical properties of the base metal using *Inventor Pro, Nastran,* and *Fusion 360* for CAD design and finite element analysis (FEA). For this phase, skewness greater than 0.75, 30°, or less for the transition angles of the mesh and parabolic tetrahedral elements for the numerical analysis was implemented, coupled with a convergence evaluation of less than 5% of error.

After the numerical evaluation, a stress path was constructed using the techniques of topological optimization, to identify the areas more susceptible to be affected if a weld was performed; leading to redesigns of geometrical features for local implementation.

Finally, an economic evaluation was made using the developed prototype as a way to compare the manufacture of the *Tie* using conventional methods and the proposed welding process, in which the size of the available machinery, tooling, and material were considered.

Results

Load Analysis

The selected component is part of the door opening mechanism of the wagons as shown in Fig. 3, where the *Tie* works as a beam for the other modules of the mechanism being attached to (Fig. 3a), allowing the displacement of the door during the operation of the train.

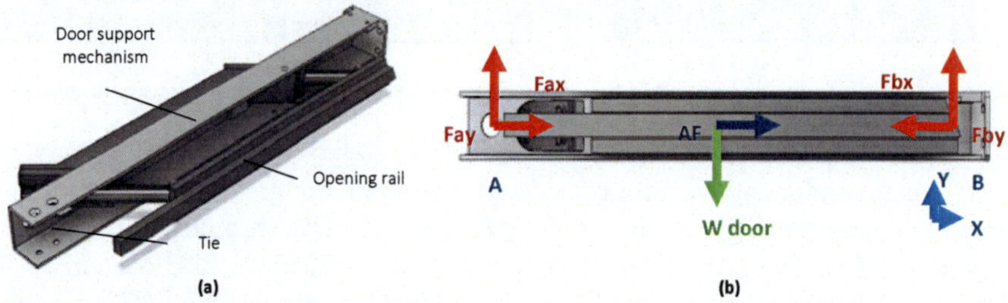

Fig. 3 Component selected—Tie. **a** Components and **b** forces. (Color figure online)

The forces applied to the *Tie* are illustrated in Fig. 3b and the considerations regarding the operation and assembly are exposed below:

- Door's weight: 35 kg (W door).
- Pneumatic actuator force at 5 bars of pressure: 2800 N (AF).
- Material used for the *Tie* AA 6082—T6.
- Material used for other components: Steel.
- Bolt connections in the holes for assembly and constraints at the end of the component (Fa and Fb).
- Two simulation conditions, when the door is fully opened and closed (74.5 mm and 22.5 mm, respectively), which are the most critical situations.

The maximum values obtained are between 62.0 and 70.8 MPa depending if the simulated condition is on the opened or closed consideration (Figs. 4 and 5). For both situations, all the locations of maximum stresses corresponded to the same regions with cracks and discontinuities as presented in Fig. 2.

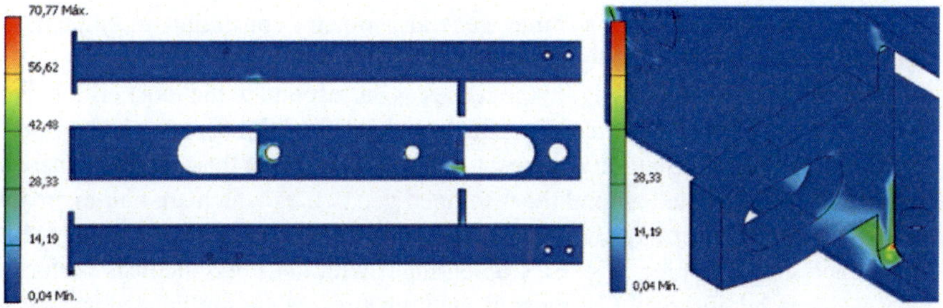

Fig. 4 Values obtained by FEA, open doors analysis. (Color figure online)

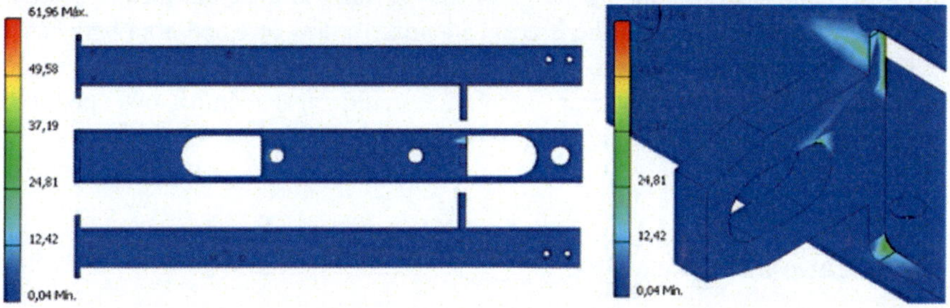

Fig. 5 Values obtained by FEA, closed doors analysis. (Color figure online)

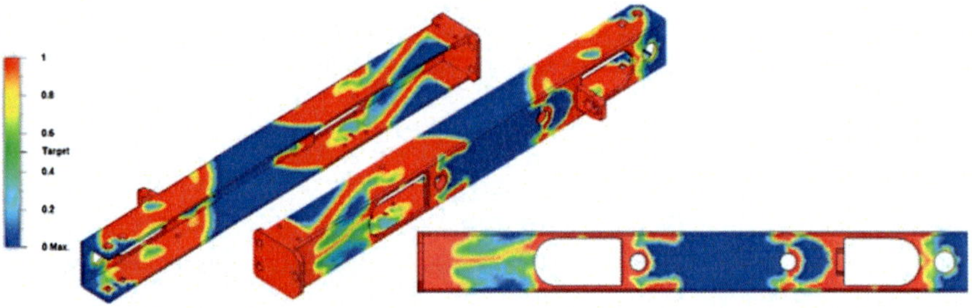

Fig. 6 Load path identification at the ties. (Color figure online)

Geometrical Considerations

The topological evaluation to identify the loading paths and useful information to be implemented in the manufacture of the *Tie* was made. Initial considerations around the areas to be preserved like bolts and contact surfaces were implemented, separated in connections at the wagon's frame, general geometry constraints to be kept, and bolt holes needed for assembly are shown.

The data is presented in Fig. 6; the red areas correspond to the most critical path, followed by the blue colour region as the least loaded material.

The stress path distribution is almost located at the edges of the geometry where the constraints were located around the machined geometries which are non-avoidable, leading to a big portion of the component to be underloaded in the middle of the piece represented in blue (Fig. 6). Comparing it with the stress analysis performed before, it can be observed that most of the high stresses are not located around the horizontal axis of symmetry (dotted line) despite the paths of loading.

This loading condition and stresses make suitable the fabrication of a longitudinal weld which makes feasible the usage of the "L"-shaped material, locally available. A detailed analysis around the bracket identifies a low stressed area between the subsequent perforation and the welded part itself, meaning that if a redesign is made to implement FSW, taking advantage of this area could translate into a lower possibility of discontinuities which could compromise the mechanical performance.

Fabrication Proposal

For each geometrical feature, a manufacturing strategy was proposed. The first is the "C" shape, despite an extruded profile presents multiple advantages, the cost involved in the fabrication of the die and the minimum volume to be produced in relation with the amount of parts needed by MdM is prohibitive. Also, a machined tie from a single block is not well suited because the volume of waste metal (around 79%) is considerable compared to a fabrication using commercial extruded parts.

The alternative selected was the usage of two L-shaped profiles and a butt joint configuration, as the sketches shown in Fig. 7a, leading to the implementation of a traditional FSW tool and traditional machinery. For the *Bracket*, three different options were evaluated (Fig. 7), where mechanical, welded, and bended approaches were compared. The construction method chosen was the machining of the complete part, followed by the fabrication of seams around the base using FSW, displacing the material with lower mechanical capabilities as farther away from the highest stresses and using the mentioned area before. This was selected because it is easier to guarantee dimensional tolerances and lesser moving parts.

To manufacture the *Lid*, an approach was made proposing different methods, and the three final alternatives are shown in Fig. 8. The selected design consists of two

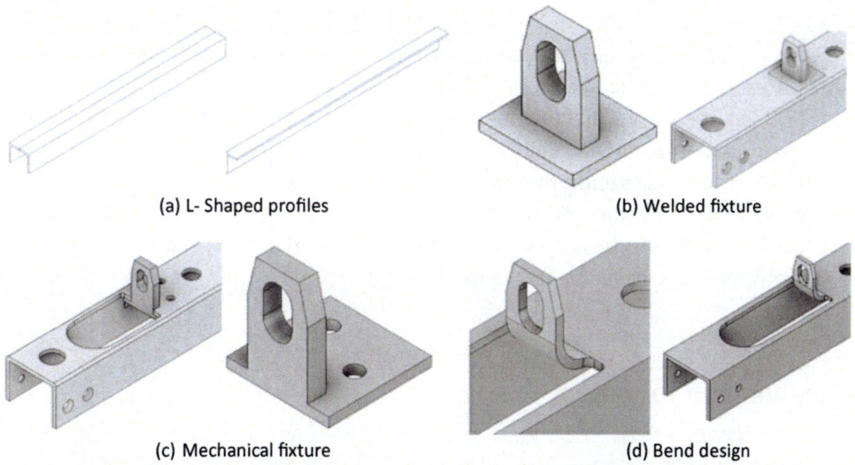

Fig. 7 **a** L-shaped profiles and configurations **b–d** Proposed bracket configurations

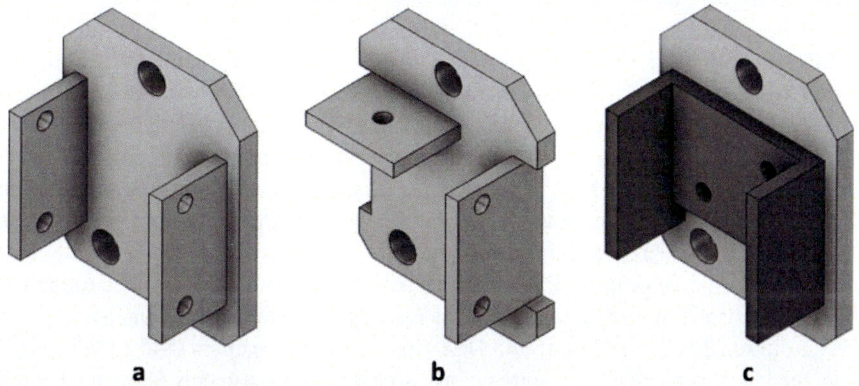

Fig. 8 Lid models for assembly, **a** Type 1 of mechanized lid, **b** Type 2 of mechanized model and **c** Modular type of lid

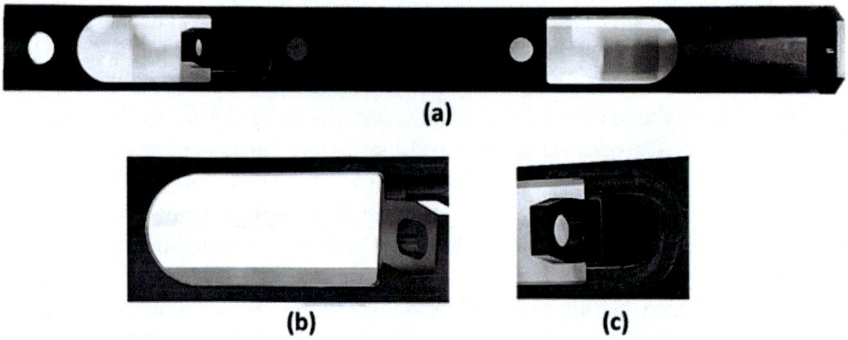

Fig. 9 Final prototype

modular pieces (Fig. 8c) which can be replaced easily and be made with another alloy to support the treads.

The final proposed manufacture is represented with a backbone weld along the length of the extrusions and another weld surrounding the bracket to join these pieces, followed by the machining required for the application.

Final images of the prototype are presented in Fig. 9, which was anodized to achieve a better look. As mentioned before, the FSW does not need any post-processing to have the great superficial finish or keep the dimensional accuracy.

Economic Evaluation

Finally, an economic evaluation was made to compare the manufacture of the component using FSW against commercially available techniques previously implemented by MdM. The base price for the analysis is the price of each local manufactured tie, around the $3'200.00 Colombian pesos (approximate 842 US dollars).

To compare the values of manufacture, Table 1 was constructed to identify the differences in price between material acquisition, machining, and FSW welding for each feature; against the locally manufactured piece. The FSW component does cost less approximately $516 USD, at present value, which translates to about 39% less money. Most of the value is related to the machining and preparation, followed by the material's cost, and the least expensive item is the FSW weld; meaning that despite the economic fluctuation of the currency, the vast majority of the price will depend on the man labour hours and the bulk material.

The alternative of using FSW does represent not only savings related to the final price but also is more efficient using the resources due to less material losses.

Additionally, the tooling ($385 USD) and fixing mechanism ($459 USD) implemented for this piece does require an investment of approximately $880, but because these components can be reused to manufacture the projected demand, the final estimated value of a single *Tie* reaches approximately $519 USD.

Table 1 Manufacturing costs

		Price (COP)	Price (US)
"C" shape manufacture	Material	$269.899	$72
	Machining and preparation	$661.170	$176
	FSW Welding	$80.672	$21
Bracket	Material	$147.390	$39
	Machining and preparation	$242.017	$64
	FSW welding	$80.672	$21
Lid	Material	$211.474	$56
	Machining and preparation	$250.664	$67
	FSW welding	–	–
Total		$1.943.960	$516

Conclusions

Friction Stir Welding is a viable alternative for repair and the replacement of low volume production of failed components in the local railway industry; the advantages of the process against more traditional welding methods do present an opportunity to broaden its applications, but the process requires a more specific analysis that could represent redesigns to the original pieces in order to fulfil the mechanical requirements.

For this particular case, the *Tie* was manufactured using the FSW process around the Bracket and along the length of the "L"-shaped extrusions, finally achieving the quality of the required dimensional tolerances.

To construct any component by FSW, a prior analysis of the geometrical characteristics must be made before any adaptation of the design due to the constraints to perform the welding technique.

Economically speaking, FSW could be cheaper to manufacture in comparison with other manufacturing techniques, for this intended purpose to be around 40% cheaper, taking into consideration the volume of production and assuming the welds can be performed by existing CNC machinery.

Acknowledgements This project was funded by RAENG under the grant agreement IAPP18-19\266. Special thanks to TWI and Metro de Medellín (MdM) for their engineering skills and support during the design process and evaluation of alternatives.

References

1. Franco F, Sánchez H, Betancourt D, Murillo O (2009) Soldadura por friccion-agitacion de aleaciones ligeras – una alternativa a nuestro alcance. Supl la Rev Latinoam Metal y Mater 1(3):1369–1375
2. Franco F, Sánchez H, Betancourt D (2012) Eficiencia mecánica en la soldadura por fricción agitación de la aleación de magnesio AZ31B Mechanical efficiency in the friction stir welding of magnesium alloy AZ31B. Redalyc 30(1):23–30
3. Escobar JD, Velásquez E, Santos TFA, Ramirez AJ, López D (2013) Improvement of cavitation erosion resistance of a duplex stainless steel through friction stir processing (FSP). Wear 297(1–2):998–1005
4. Pérez SP, Insignares IM, Charris CO, Posada J, Silgado JU (2014) Medición Del Torque Durante La Soldadura Por Fricción-Agitación De Aluminio Mediante Un Sistema De Detección Con Transmisión En Tiempo Real. Rev Colomb Mater 5:244–249
5. Rodríguez RJ, Caballis CA, Cely BMM, Unfried-Silgado J (2018) A comparative study of corrosion resistance in welded joints of aluminium alloy AA1100 obtained by friction-stir and gas metal arc welding processes|Estudio comparativo de la resistencia a la corrosión en juntas soldadas por fricción-agitación y por el. Ingeniare 26(3):419–429
6. Hoyos E, Montoya Y, Wilches LV, García D, Castaño V (2020) DISEÑO DE UNA HERRAMIENTA DE HOMBRO ESTACIONARIO PARA LA OBTENCIÓN DE JUNTAS DISIMILES ALUMINIO - POLÍMERO POR FSW, Envigado
7. Kwon YW, Bang H (1997) The finite element method using MATLAB. CRC Press, Minnesota
8. Kumagai M, Tanaka S (1999) Properties of aluminum wide panels by friction stir welding. In: 1st international symposium on FSW
9. Ohba H, Ueda C, Agatsuma K (2001) Innovative vehicle - the 'A-train'. Hitachi Rev 50(4):130–133
10. Blach O, Senne F (2002) Reibrührschweißen aus der Sicht eines Anwenders im Schienenfahrzeugbau, 2nd GKSS Work
11. TWI (2020) Friction stir welding on the London underground - TWI. https://www.twi-global.com/media-and-events/insights/friction-stir-welding-on-the-london-underground-383. Accessed 18 Aug 2020
12. MatWeb (2020) Aluminum 6082-T6. Aluminum Alloy. http://www.matweb.com/search/DataSheet.aspx?MatGUID=fad29be6e64d4e95a241690f1f6e1eb7. Accessed 25 Aug 2020
13. MatWeb (2020) Aluminum 6063-T6. Aluminum Alloy. http://www.matweb.com/search/DataSheet.aspx?MatGUID=333b3a557aeb49b2b17266558e5d0dc0&ckck=1. Accessed 25 Aug 2020
14. MatWeb (2020) Aluminum 7075-T6; 7075-T651. Aluminum Alloy. http://www.matweb.com/search/DataSheet.aspx?MatGUID=4f19a42be94546b686bbf43f79c51b7d. Accessed 25 Aug 2020

Three-Sheet Al Alloy Assembly for Automotive Application

Piyush Upadhyay, Hrishikesh Das, and Daniel Graff

Abstract An update on the joint characterization of three-sheet friction stir lap welds (FSLW) in three aluminum alloys (7xxx, 5xxx, and 6xxx) for automotive assembly application is presented. Welding speed of 0.3–2.9 m/min was used with pin length >5 mm with the aim to reduce upturn hooking on the retreating side and avoid wormhole defect on the advancing side. Joints were also made to enable T peel and U peel testing. Stitch welds were used to understand the effects of weld length and ramp length on T peel performance. Hat sections were joined for three-point bending and axial crush testing and tested with promising results. Several key issues that require further investigations are identified including flash management near the trim edge, weld quality with limited clamping, tool failures at speeds greater than 0.5 m/min, and the propensity of incipient melting for 7xxx stack-up.

Keywords Friction stir lap welding · High speed · AA7xxx · AA6xxx

Background and Introduction

Friction stir lap welding has emerged as an enabling technology to assemble similar and dissimilar Al alloys [1–5]. A number of research work are reported on the microstructure and mechanical property correlation, optimization of process parameters, and effect of tool geometries for butt welding of aluminum alloys [1–5]. Work on friction stir lap welding (FSLW) of similar and dissimilar aluminum alloys are scarce. Unlike butt joining, FSLW interface is oriented perpendicular to the axis of tool rotation. Consequently, FSLW has a tendency for inadequate mixing during multi-sheet stack-ups. Furthermore, undispersed surface oxide at the interface can lead to undesirable material upturn on either side of the nugget, often referred to as "hooks." Additionally, the bonded area for FSLW is more dependent on pin diameter rather than the much larger shoulder diameter as is the case in butt joining; thus, material flow around pin (i.e. faying interface) is critical for FSLW. Lap geometry

P. Upadhyay (✉) · H. Das · D. Graff
Applied Materials and Manufacturing Group, Pacific Northwest National Laboratory, Battelle Boulevard, Richland, WA 99354, USA
e-mail: Piyush.Upadhyay@pnnl.gov

Y. Hovanski et al. (eds.), *Friction Stir Welding and Processing XI*, The Minerals, Metals & Materials Series, https://doi.org/10.1007/978-3-030-65265-4_2

inherently contains crack-like structures at both the edges of the overlap. In many cases, this hook feature can either deviate towards top or bottom of the joint, reducing the effective sheet thickness concurrently acting as a site for crack initiation and propagation, thus adversely affecting mechanical properties of the joint. Management of this interface is key to obtaining acceptable static and dynamic joint performance.

While first-order estimates on how weld parameters can affect joint morphology are understood, functional relationships between control parameters, tool design, and the resulting interface that are critical to joint strength remain elusive. It is observed that in some instances the presence of hook features is especially detrimental when on the advancing side of the joint. The problem is exacerbated when the hook is deviated towards the sheet with a lower load-carrying capacity.

Certain combinations of welding control parameters and tool geometries have been shown to produce superior welds in the literature; however, these parameters are highly specific to a chosen material stack-up. The reported joint strength values have a large scatter and can range anywhere between 20 and 60% of the base metal. Additionally, the welding speeds reported in the literature are relatively low (most of the welds are performed at around 0.3–0.6 m/min while only a few show capabilities up to 1 m/min See Fig. 1) and do not justify commercial investment nor a switch to FSLW technology unless a significant improvement in welding speed can be demonstrated.

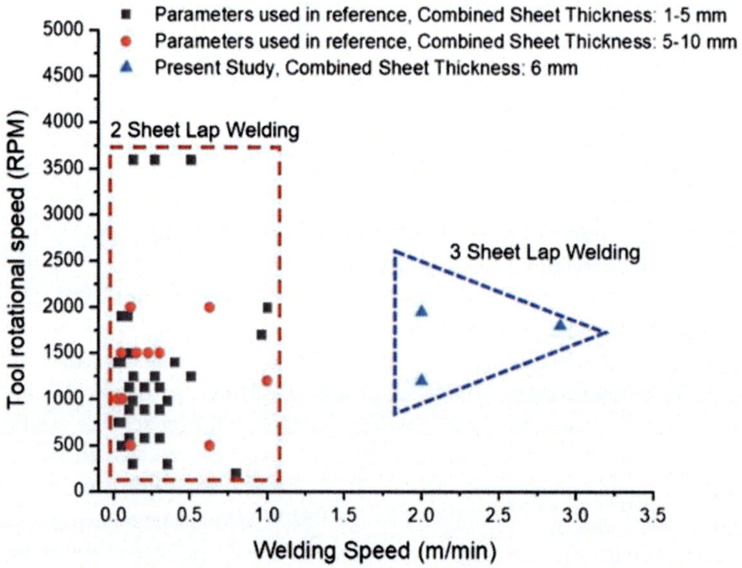

Fig. 1 Current status of welding parameters (welding speed and tool rotation speed) and thickness studied for two-sheet FSLW aluminum alloys. Data from previously published research works are indicated by red circle and black rectangle and present work with blue triangles. (Color figure online)

Pacific Northwest National Laboratory (PNNL) in collaboration with an original equipment manufacturer and a material supplier is focused on evaluating and developing three-sheet FSLW such that Al alloys relevant to automotive assembly line can be effectively lap welded. In 2019, FSW&P proceeding, an overview of process development and tool design effects on joint lap shear performance in three stack up material set was provided. In this subsequent paper, we provide an update on progress made, and different testing schemes adopted to understand the joint performance with focus on T peel testing.

Materials and Experimental Details

Welds were made with lap stack of 7055 (2.5 mm) – 7055 (2.5 mm) – 6022 (1 mm). FSW tools were made of heat-treated H-13. Joints were produced on a high precision Manufacturing Technology, Inc., gantry-type FSW machine at PNNL. The FSW system can measure several process responses in real time including tool forces in all three directions, tool torque, and position.

T peel samples were obtained from pre-bent U-shaped joints. Typically used to qualify spot welded and adhesive joints KSII samples employ symmetric loading to test similar or dissimilar joints. To the author's knowledge, this is the first use of KSII setup in Friction stir lap welded joints. A tailored clamping and fixture arrangement were designed and set up to U-shaped samples for KSII testing (See Fig. 2). While KSII testing was done for various configurations [5], in this paper, only T peel samples extracted from KSII configurations are discussed. In addition, unguided two stitch weld lap shear tensile tests were also performed and reported.

Fig. 2 Schematic and experimental welding and testing set up for 3-sheet KSII configuration. (Color figure online)

Results and Discussion

The effect of long versus short stitch weld can be appreciated from T peel test results as shown in Fig. 3. T peel coupon containing two stitches shows higher failure load compared to 1 stitch weld with same total weld length. Two stitches provide two plunge-in region where the material mixing is expected to be better. Additionally, the gap between the two stitches creates a barrier to crack propagation thus increasing the overall load-bearing capacity.

To obtain further clarity on the effect of length of a continuous weld on load-bearing capacity, a unit force study was performed for stitch welds ranging from 10 to 50 mm. Four different welding schemes with varying ramp-up to full welding speed and weld length were used to produce T peel samples. Representative load versus extension plots for each case are shown in Fig. 4. The corresponding welding speed versus welding distance used for each run is shown in the inset.

While the absolute value of load-bearing capacity increases with the increase in the length of the weld, the corresponding increase in the load-bearing capacity is not linear. The shortest stitch weld exhibited the highest load per unit weld length and joint stiffness values. A load-bearing capacity ranging from 200 N/mm (for 10 mm stitch) to 125 N/mm (for 50 mm stitch) was observed. With the same weld length of 30 mm joint with a longer ramp performed marginally better (Fig. 4b vs. c). Both 30 mm welds failed in an analogous manner from the top sheet HAZ. 50 mm weld failed by tearing the bottom sheet. However, 10 mm weld fractured through the HAZ of the top sheet.

In addition to transverse T peel, testing was also performed to assess in the longitudinal configuration. Two stitch longitudinal T peel samples were welded in KSII configuration and extracted for T peel testing (Fig. 5). Samples were tested in two-loading direction: weld exit-hole loaded on the top sheet and weld start loaded on

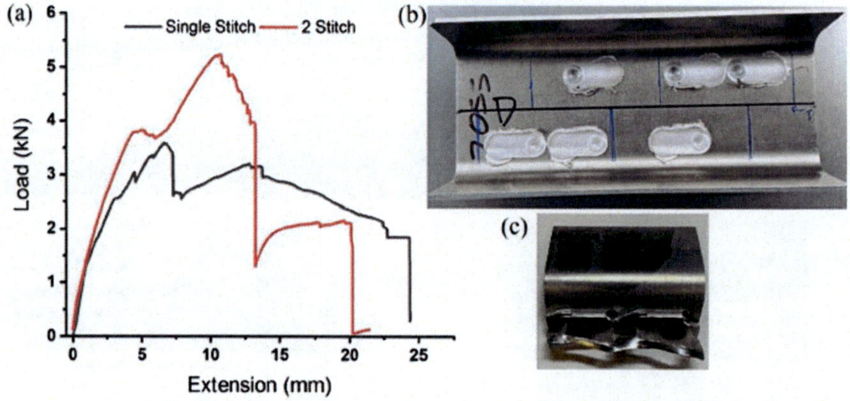

Fig. 3 Comparison of 1 and 2 stitch of 20 mm each **a** T Peel strength in longitudinal direction, **b** 1 and 2 stitch welded sample in KSII configuration, and **c** fracture mode for 2 stitch welds. (Color figure online)

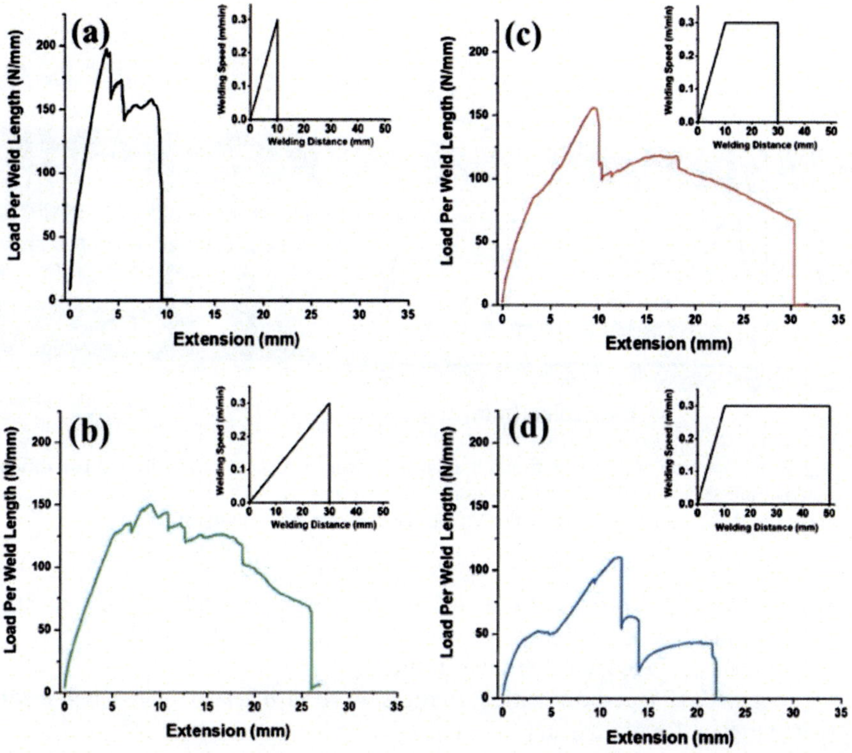

Fig. 4 Unit force T peel test data set for **a** 10 mm ramp, **b** 10 mm ramp + 20 mm constant RPM = 30 mm, **c** 10 mm ramp + 40 mm constant RPM = 50 mm, **d** 30 mm ramp, weld length. (Color figure online)

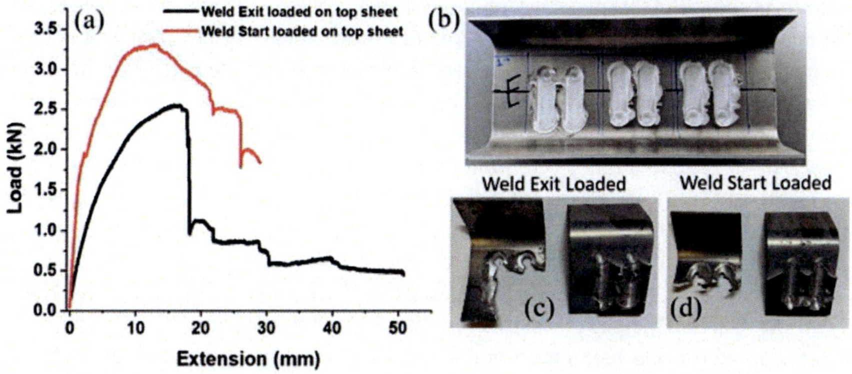

Fig. 5 Comparison of **a** longitudinal T Peel strength for weld exit loaded on top sheet and weld start loaded on top sheet, **b** longitudinal two stitch weld in KSII configuration, **c** failure mode for exit-hole loaded on top sheet **d** failure mode for plunge-in/start loaded on top sheet. (Color figure online)

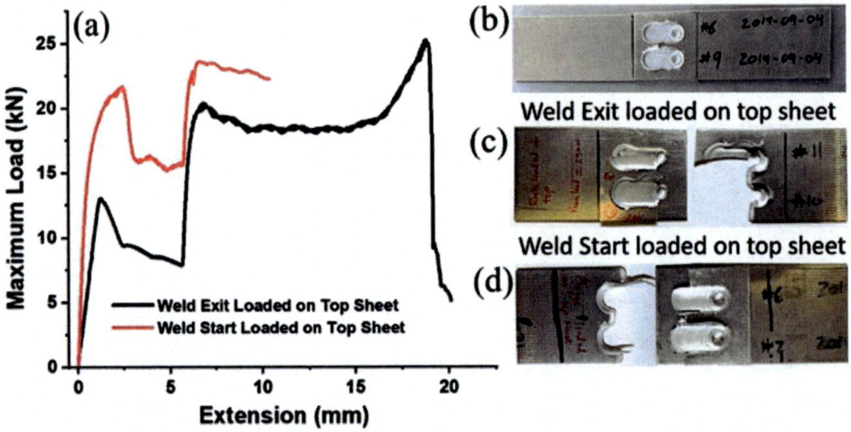

Fig. 6 Comparison of 2 × 20 mm stitch **a** lap shear tensile strength comparison in two different loading direction, **b** lap shear welded sample, failure mode for sample **c** exit-hole loaded on top sheet and **d** plunge-in/weld start loaded on top sheet. (Color figure online)

the top sheet. Weld start loaded on top sheet shows higher failure load compare to weld exit loaded. This suggests that greater mixing of the top two sheets expected at the weld start resulted in better performance.

Longitudinal lap shear tensile samples were also tested in the two-loading configuration. Results are shown in Fig. 6.

Surprisingly, exit-hole loaded on top joint showed slightly greater load-bearing capacity compared to start loaded sample. The load curve and fracture mode appear different for the two cases. Start loaded sample shows lower absorbed energy (area under the curve) and the sample fractured by tearing from the bottom sheet plunge-in location with almost intact the remainder of the weld stitches. Whereas, exit-hole loaded lap shear sample exhibited greater absorbed energy (area under the curve). The fracture mode is complex compared to plunge direction loading. For this sample, crack initiates through the HAZ of the first stitch. Crack path then changes direction around the weld nugget moving towards the second stitch. This leads to a secondary peak in the load and finally the joint pulls out from the exit-hole of the top sheet.

This unit study of load-bearing capacity provides a clearer picture of the expected mechanical performance in FSLW joints that can enable informed engineering design of aluminum assembly including weld path and stitch design.

Subsequently, we successfully demonstrated three-sheet FSLW on Hat-Hat-Hat and Hat-Hat Flat flange section for crush and 3-point bend testing (Fig. 7a). Moreover, the adopted process parameters and setup used are quite stable for long welds with repetitive runs. As an example, force response for two side seams (Fig. 7a) in a narrow flange are shown in Fig. 7b. All the 3 Force (X, Y, and Z) response provide repeatable performance.

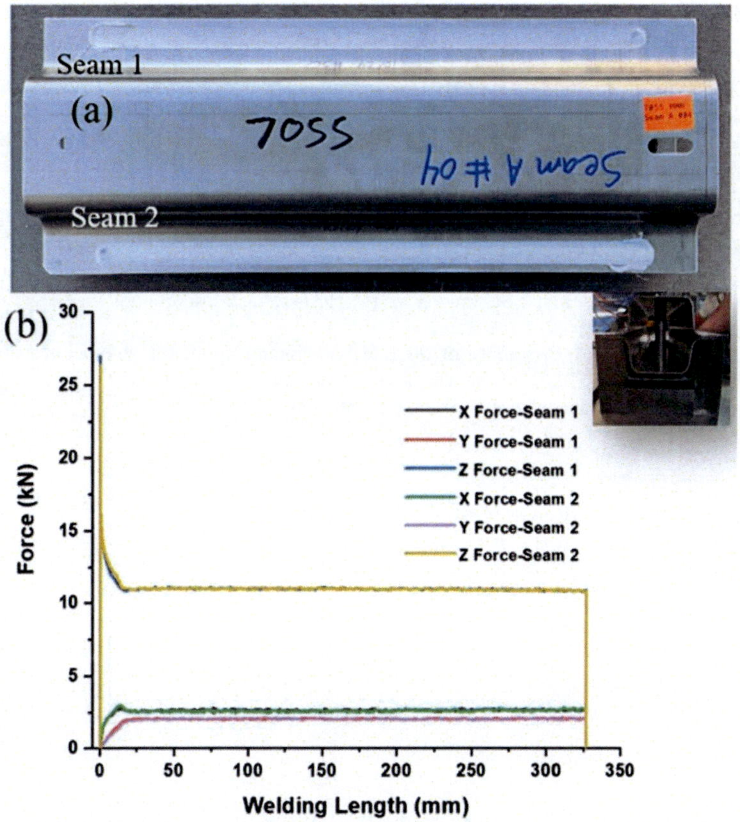

Fig. 7 **a** Successful implementation of high speed three-sheet FSLW on Hat-Hat-Hat Flange section, **b** side view of Hat-Hat-Hat section. (Color figure online)

Summary

The goal of this project is to mature friction stir lap welding technique for three-sheet Al stacks up such that joints can be made at industrially viable welding speed that meet industrially relevant design criteria. By studying a wide variety of welding parameters and fabricating and testing sample in different loading conditions we have gathered a large volume of relevant information to assist in component design for Al assembly.

Acknowledgements PNNL is operated by Battelle Memorial Institute for the U.S. Department of Energy under contract DE-AC05-76RL01830. This work was funded under US DOE Technology Commercialization Fund under office of Technology Transition and Vehicle Technologies office with contributions from participating entities including Honda R &D Americas LLC, and Arconic Inc. The authors acknowledge input and guidance from Eric Boettcher, HRA for KSII and unit force study. The authors also acknowledge assistance of Tim Roosendaal and Ortiz Angel for sample preparation and testing.

References

1. Cederqvist L, Reynolds AP (2001) Factors affecting the properties of friction stir welded aluminum lap joints. Weld J-N Y 80(12):281-s
2. Song Y, Yang X, Cui L, Hou X, Shen Z, Xu Y (2014) Defect features and mechanical properties of friction stir lap welded dissimilar AA2024–AA7075 aluminum alloy sheets. Mater Des 55:9–18. https://doi.org/10.1016/j.matdes.2013.09.062
3. Threadgill PL, Leonard AJ, Shercliff HR, Withers PJ (2009) Friction stir welding of aluminium alloys. Int Mater Rev 54(2):49–93
4. Upadhyay P, Reynolds AP (2010) Effects of thermal boundary conditions in friction stir welded AA7050-T7 sheets. Mater Sci Eng: A 527(6):1537–1543. https://doi.org/10.1016/j.msea.2009.10.039
5. Upadhyay P (2020) Assembly of dissimilar aluminum alloys for automotive applications. Presented at the VTO/DOE annual merit review meeting, 1 June 2020. https://www.energy.gov/eere/vehicles/annual-merit-review-presentations

Characterization and Analysis of Effective Wear Mechanisms on FSW Tools

Michael Hasieber, Michael Grätzel, and Jean Pierre Bergmann

Abstract This study systematically analyzes effective wear mechanisms at the shoulder and probe, separately for plunging and welding. The investigations were carried out with a robotic welding setup in which AA-6060 T66 sheets with a thickness of 8 mm were joined by weld seam lengths of up to 80 m. To compare and differentiate the wear mechanisms between plunging and welding, repeated plunging cycles were initially investigated without tool movement. Subsequent experiments consider welding as well, whereby adhesion, abrasion, surface fatigue, and tribochemical reactions which were characterized for various weld seam lengths. During welding of the AA-6060 T66 sheets, the tool material 1.2344 (X40CrMoV5-1) exhibited abrasive wear that occurred due to self-damaging of the FSW tool. The wear analysis showed that no significant wear occurred during the plunging stage.

Keywords Friction stir welding · Tool wear · FSW · Wear characterization · Wear mechanisms · Plunging · Welding · 1.2344 · AA-6060 T66

Introduction

The increasing use of aluminum alloys and rising industrial requirements is driving a need for the development of appropriate welding processes. Against this background, friction stir welding (FSW) is a promising solid-state joining technique, mainly used for the welding of aluminum alloys. In contrast to conventional joining methods such as gas tungsten arc welding (GTAW), the advantages of this technique include the absence of consumables (cover gases and filler materials) and the comparatively low joining temperature, which prevents the formation of pores and hot cracks [1]. Due to its excellent mechanical weld seam properties, FSW has been implemented in automotive [2], shipbuilding [3], aerospace [4], and other applications.

In general, the functional element of the FSW process is a rotating tool consisting of a shoulder and probe (see Fig. 1). During welding, the tool plunges into the

M. Hasieber (✉) · M. Grätzel · J. P. Bergmann
Department of Manufacturing Technology, TU Ilmenau University, 98693 Ilmenau, Germany
e-mail: michael.hasieber@tu-ilmenau.de

Y. Hovanski et al. (eds.), *Friction Stir Welding and Processing XI*, The Minerals, Metals & Materials Series, https://doi.org/10.1007/978-3-030-65265-4_3

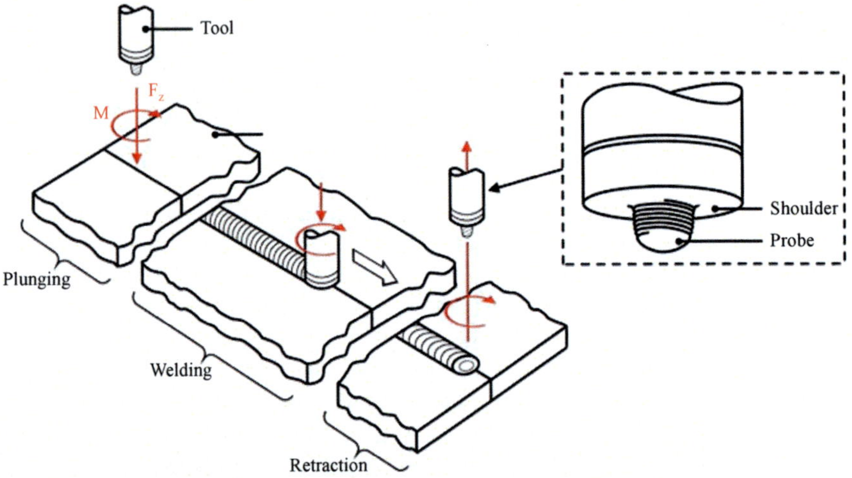

Fig. 1 Principle of FSW in accordance with [7]. (Color figure online)

workpiece causing the workpiece material to be plasticized by frictional heat. The complex, induced material flow interactions, which form the weld seam behind the tool [5, 6], are affected by the shoulder and probe geometry, rotational speed, and lateral movement. Figure 1 depicts the working principle of the FSW process, including the plunging, welding, and retraction phases.

During plunging, the rotating tool is pressed into the joining area with an axial force F_Z. The plunging stops when the shoulder comes into contact with the surface of the workpiece. A defined dwell time can be set to sufficiently plasticize the workpiece material. After plunging and dwelling, welding occurs. During welding, the FSW tool is moved along the joint surfaces and the plasticized workpiece material is continuously extruded and displaced by the shoulder and the probe while below their liquidus temperatures. After welding, the tool is retracted from the joining area [8]. However, FSW is subjected to process-specific challenges, including comparatively high process forces, low welding speeds, and FSW tool wear resulting from tribological interaction between the tool and workpiece. The geometric-related FSW tool wear causes limitations in the process stability, including the risk of premature failure [9], lateral path deviations and varying material flow [10]. The detection and analysis of the quantitative and qualitative tool wear is thus essential for improvement of tool life, consistent weld seam quality, and the reduction of irregularities. This requires a deeper understanding of how the process phases influence the resulting wear and what type of wear mechanisms occur. Against this background, a disaggregated identification of the wear mechanisms on shoulder and probe were carried out as a function of plunging and welding.

Several studies have investigated how the geometric deviations on shoulder and probe depend on tool material, joining material, and process parameters. For example, wear investigations were carried out on aluminum alloys as the joining material with tools made of tool steel (O1 and H13) [11], on steel alloys with tools made of tungsten–rhenium (W–Re) [12, 13], and on titanium alloys with tools made of tungsten carbide–cobalt (WC–Co) [14, 15]; furthermore, it could be shown that wear increases at high energy input due to the process parameters: high rotational speeds and low welding speeds [16–19]. In contrast, the influence of the process phases on tool wear is scarcely considered.

Only theoretical assumptions on increased tool wear during plunging are reported in Prado et al. and Weinberger et al. The increased wear is explained by the fact that the tool plunges into the cold material under increased loads [17, 20]. To avoid tool wear during plunging, Sahlot et al. used pilot holes with the same dimensions as the probe. The wear behavior of copper–chromium–zirconium (CuCrZr) alloy with tools manufactured of H13 tool steel (hot-working steel X40CrMoV5-1) was examined during FSW and exhibits significant wear at shoulder and probe for rotational speeds up to 1200 rpm [21]. An analysis of the wear resulting from the comparative plunging was not carried out. In this context, only comparisons with friction stir spot welding (FSSW) are available to date [20]. The investigations of Choi et al. represent tool wear as a function of up to 500 plunging procedures. Tool wear was examined during FSSW of low carbon steel plates (0.6 mm thickness) with tools manufactured from tungsten carbide–cobalt (WC − Co). The results show that the probe length was significantly changed after 100 plunging procedures due to wear [22].

However, it should be noted that the results of FSSW are only conditionally meaningful because the tool geometry differs with FSW. In addition, FSW causes thermal and mechanical stresses at the transition between plunging and welding that have a significant impact on tool wear. In addition to the consideration of wear for the respective process phase, the determination of wear mechanisms is essential for a comprehensive analysis. Therefore, a general description of the wear mechanisms within a tribological system is necessary. The interaction within a tribological system of base body and counter body results in different wear mechanisms. These can be classified as adhesion, abrasion, surface fatigue, and tribochemical reactions [23]. Adhesion occurs under tribological stress as a result of high local pressures on certain surface irregularities, resulting in local interfacial bonds with high strength. During a relative movement of base body and counter body, separation occurs in the surrounding base material but not between the interfacial bonds [23, 24]. In tribological contacts, abrasion occurs if the counter body is considerably harder than the base body or if hard particles penetrate the material. As a consequence of relative movement, abrasive wear can occur in the softer body through various material separation processes (micro-grooving, micro-chipping, and micro-fracturing). Surface fatigue appears during alternating mechanical and thermal stress. Microstructural changes, fatigue, crack formation, and crack growth lead to removal of wear particles and to larger material breakouts [23]. Tribochemical reactions are chemical reactions of the body and counter body within a tribological system due to diffusion processes. The tribologically stressed surfaces interact with the surrounding material so that new reaction products are constantly generated and worn out during relative movement [23, 24].

After presentation of the wear mechanisms, the wear mechanisms are classified within the context of the FSW using the process phases as shown in Fig. 1. The aggregative wear analysis for plunging and welding as a function of joint quality and weld seam length was carried out by Wieckowski et al. During FSW of conventional aluminum alloy AA-7075 T6 with tools made of H13 tool steel and MP-159 alloy (Ni-Co), geometric deviations and material adherence at probe and shoulder were observed [11]. Thompson et al. and Hoßfeld et al. characterize the material adherence caused by the joining of conventional aluminum alloys as an indicator of adhesive wear [12, 25]. Furthermore, the investigation of Tarasov et al. shows the formation of $FeAl_3$ reaction layers during FSW of conventional aluminum alloy (AMg5M) with tools manufactured of H13 tool steel due to tribochemical reaction [26]. As a result of hard material particles embedded in the matrix material, both adhesion and abrasion are main wear mechanisms for joining aluminum MMCs (metal-matrix composites) [1, 17, 20]. To evaluate the effectiveness of harder tool materials, Prater et al. exhibits the wear in FSW of cast Al359 + 20% SiC and Al359 + 30% SiC matrix composites with tools made of O1 tool steel, cemented carbide (WC–Co) of both micrograin and sub-micrograin varieties and WC–Co coated with diamond. It was demonstrated that while harder tool material can extend tool life [27], they are also more brittle and tend to fail prematurely [17]. Results such as brittle fracture and crack initiation during FSW are shown from Tiwari et al. for shipbuilding steel DH36 with tools of different grades of tungsten carbide (WC-6 wt.% Co, WC-10 wt.% Co) [28].

In summary, numerous studies have focused mainly on the wear mechanisms in FSW for higher melting materials such as steel. Furthermore, the results for FSW tool wear usually aggregated plunging and welding. The aim of this paper is to systematically analysis the effective wear mechanisms at shoulder and probe, disaggregating the plunging and welding processes. The analysis and comparison of plunging and welding enables the contribution to wear from each process phase and their significance to be determined. From this determination of wear mechanisms, it is possible to initiate demand-adapted improvements to increase tool life. In addition, identifying the separate contributions of plunging and welding to wear allows the adaptation of process controls to minimise wear within the respective phases.

Experimental Procedure

The tests were carried out on the widely used aluminum alloy AA-6060 T66 with a length of 1000 mm, a width of 50 mm, and a thickness of 8 mm. The shoulder and probe were manufactured from hot-working steel (X40CrMoV5-1) with a subsequent hardening procedure. The Rockwell hardness of the probe and shoulder are about 50 h and 60 h, respectively. The chemical composition of the workpiece and tool material is shown in Tables 1 and 2.

As can be seen in Fig. 2, the probe has a diameter of 9 mm and a tapered base body. To increase the material flow, the probe was designed with a thread and three flanks displaced by 120° to each other. The length of the probe was adjusted to

Table 1 Chemical composition of AA-6060

Element	Si	Fe	Cu	Mn	Mg	Cr	Zn	Ti	Al
Min. %	0.3	0.1	–	–	0.35	–	–	–	Bal
Max. %	0.6	0.3	0.1	0.1	0.6	0.05	0.15	0.1	

Table 2 Chemical composition of X40CrMoV5-1

Element	C	Si	Mn	P	S	Cr	Mo	V	Fe
Min. %	0.35	0.8	0.25	–	–	4.8	1.2	0,85	Bal
Max. %	0.42	1.2	0.5	0.03	0.02	5.5	1.5	1.15	

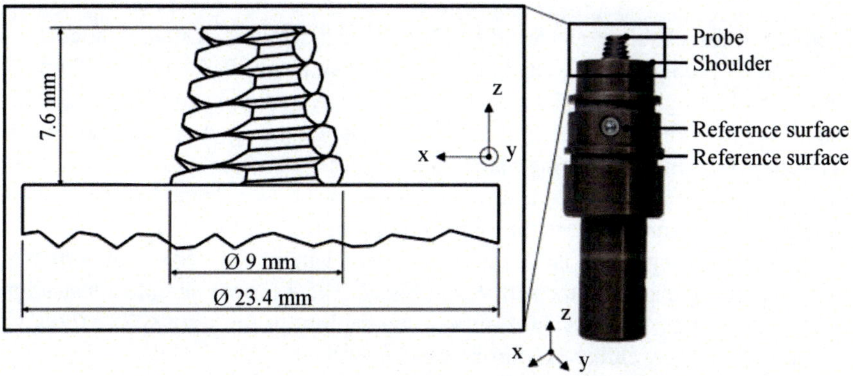

Fig. 2 Dimensions of the FSW tool with the measuring plane in accordance with [10]

7.6 mm, which is 95% of the sheet thickness. The shoulder was concave and has a diameter of 23.4 mm. The wear examinations were performed with a rotational speed of 3333 rpm, an axial force of 7.29 kN, and a feed rate of 200 mm/min.

Tool wear was measured in repeated welding tests, separately for plunging and welding. The experiments were repeated three times for statistical robustness. The tests were performed with a force-controlled robotic FSW system from Grenzebach Maschinenbau GmbH using a KUKA KR 500 MT robot. After the welding tests, the wear on shoulder and probe was assessed by visual inspection and optical measurement using stripe light projection (GOM ATOS Core 300). Deviations of the geometry were detected by the shapes of the projected light sections arising from wear on shoulder and probe. This allows the local determination of the tool wear in three dimensions by combining several partial images in a digital model. In addition to geometric deviations, volume deficit and thus the weight loss can be determined. The maximum measuring range of the optical 3D scanner is 300 mm with a measuring accuracy of 0.02 mm (Fig. 3).

The wear investigations on shoulder and probe started with a reference measurement using the initial state of the tool. Subsequently, the development of wear was

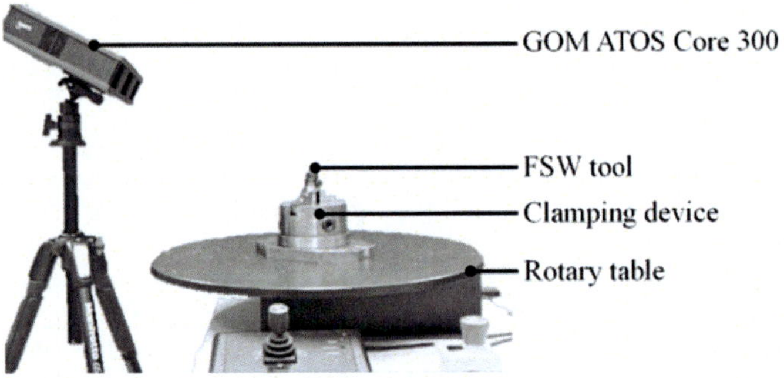

Fig. 3 Measuring system with the optical scanner GOM ATOS Core 300 in accordance with [10]

examined for plunging and welding. The characterization was taken after 21, 42, 63, and 84 plunging procedures and welding lengths of 20, 40, 60, and 80 m. The number of plunging procedures results from the maximum weld seam length of the clamping fixture of 0.95 m (x-direction), as shown in Fig. 4. Based on a welding distance of 20 m, 21 plunging procedures are necessary.

As in previous work [10], a measurement plane was defined to prevent false measurements and to ensure the comparability of the results, see Fig. 5.

The measuring plane is located in the x-z-plane of the FSW tool, which was oriented using the reference surfaces in Fig. 2. The wear measurement was aligned to the measuring plane to ensure the comparability of the results. The tools were cleaned after the welding experiments in a solution of 25% NaOH.

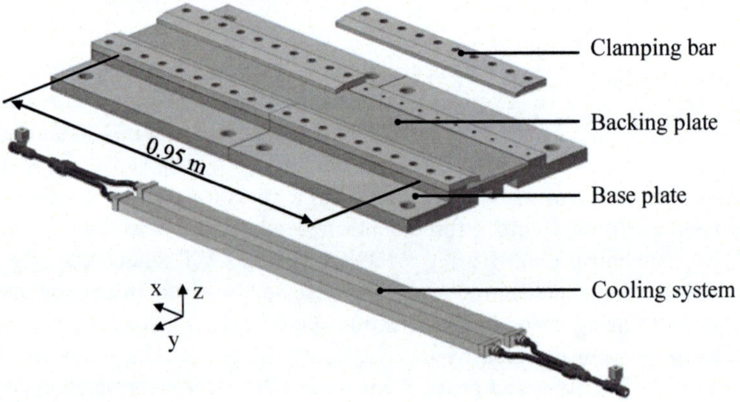

Fig. 4 Clamping fixture

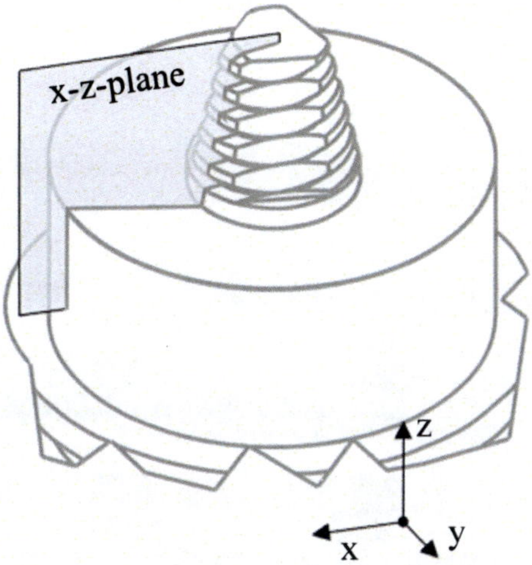

Fig. 5 Measuring plane on the FSW tool in accordance with [10]

Results and Discussion

Previous studies by Hasieber et al. examine the determination of the geometric- and weight-related tool wear for the same shoulder and probe geometry used in the present study [10]. The qualitative tool wear characterization, shown in Fig. 6, was carried out using photographic images and stripe light projection to identify surface deviations on the probe. The measuring plane shown in Fig. 5 was used to ensure equal alignment of the FSW tools during the qualitative tool wear characterization.

Figure 6a shows FSW tool wear for varying plunging procedures (rows) and measurement methods (columns) in the y–z-plane. The repeated characterization was performed to ensure comparability with the aggregative consideration after 21, 42, 63, and 84 plunging procedures. The changeover time between the individual plungings was 10 s. The number of plunging procedures results from the maximum weld seam length of the clamping fixture of 0.95 m, as shown in Fig. 4. Based on a welding distance of 20 m, 21 plunging procedures are necessary. Both the photographic images and the light stripe projection reveal no significant wear deviations on the probe relative to the initial state for an increasing number of plunging procedures. A change in the shape of structural elements such as threads or flanks of the tapered probe were not identified either. On the other hand, Fig. 6b shows a significant change in shape for increased plunging procedures and the corresponding weld seam lengths after 20, 40, 60, and 80 m. As already shown in previous work [10], marginal form deviations occur on geometrical structures of the probe between 0 and 20 m. In addition, a smoothing of the edges between the major thread radius and the thread flanks due to wear was identified. Furthermore, it was shown in [10] that an increase of the weld seam length up to 80 m leads to significant deviations and partial breakouts in the major thread radius and thread depth. Three-dimensional measurement was used to illustrate the effect of the local wear on shoulder and probe of varying plunging

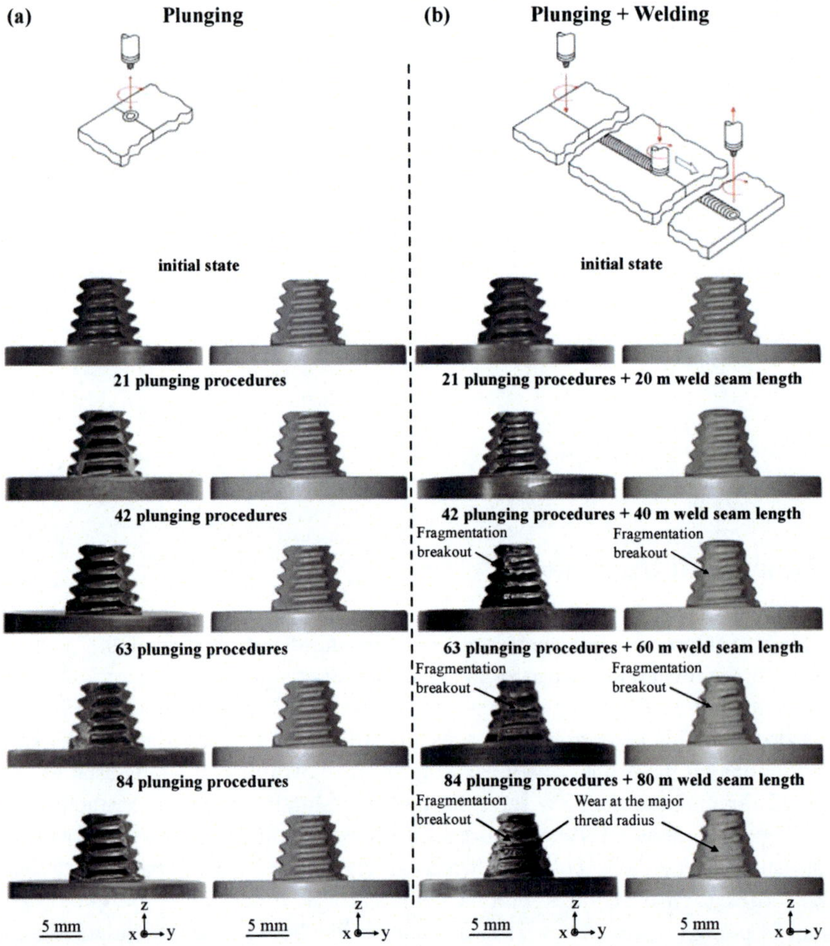

Fig. 6 Qualitative inspection for varying plunging procedures (**a**) and weld seam lengths (**b**) using photographic images and stripe light projection in accordance with [10]. (Color figure online)

procedures and the corresponding weld seam lengths of up to 80 m. As in Hasieber et al., the digital models from the qualitative inspection were compared with the initial state to quantify the wear on shoulder and probe, see Fig. 7.

The colored scale in Fig. 7 symbolizes FSW tool wear, where blue indicates no wear and red indicates significant wear. Figure 7a depicts the surface measurements for up to 84 plunging procedures. Marginal wear at the transition zone of thread and flank of the tapered probe relative to the initial state was found between 21 and 63 plunging procedures. However, the wear was only observed on the side of the flank facing the direction of rotation. After 84 plunging procedures, slight wear was visible on the thread flanks that face in the direction of the weld root and on the truncated cone of the probe. Compared to welding, repeated plunging exhibited no

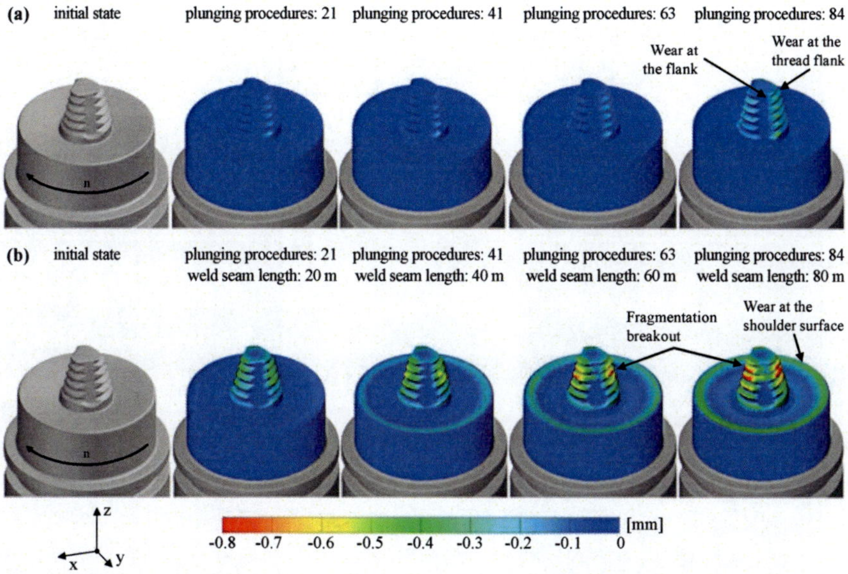

Fig. 7 3D surface measurement for **a** 84 plunging procedures and **b** weld seam lengths of up to 80 m using stripe light projection in accordance with [10]. (Color figure online)

significant wear deviations on the probe and shoulder. A plausible explanation for this behavior is the contact time between the FSW tool and the workpiece material, which is significantly shorter during plunging than during welding. In addition, wear is also dependent on the temperature [23]. While the temperature increases during plunging, a high and almost constant temperature prevails during welding. The results from Fig. 7b refer to previous work [10] and show that FSW tool wear is mainly induced through welding. Massive wear has been identified on the probe and the shoulder, leading to significant geometrical deviations. A detailed consideration of the disaggregated wear behavior in FSW is described in [10].

After identifying areas with significant wear, a characterization of the wear mechanisms at shoulder and probe was carried out using scanning electron microscopy (SEM). The investigation of the wear mechanisms started with a detailed reference measurement of the tool in the initial state, see Fig. 8.

Figure 8a depicts a partial view of the probe at a magnification of 30:1. At a magnification of 250:1 and 2000:1, dark image areas are visible on the main thread radius. It can be assumed that these are carbides or carbide clusters, formed during the tempering process and ensure high-temperature strength and high-temperature hardness of the tool material. The determination of these is subject to further research and can provide a further approach to a deeper understanding of tool damage. At a magnification of 30:1, the partial view of the shoulder, as shown in Fig. 8b, exhibits no noticeable features except manufacturing traces from the turning process. On closer examination, dark image areas similar to the probe are visible. The reference

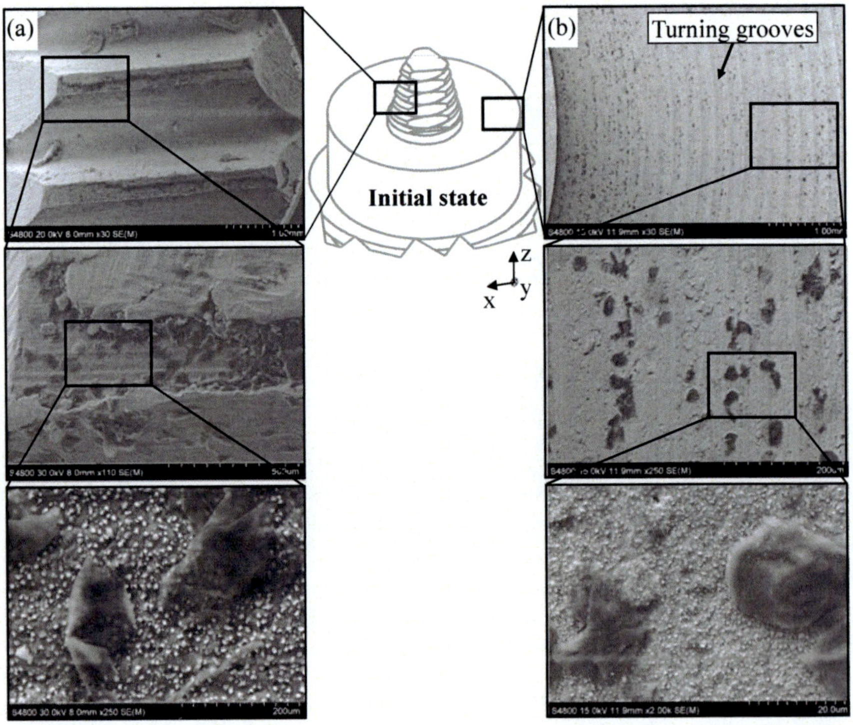

Fig. 8 SEM analysis of the FSW tool in the initial state at the probe (**a**) and the shoulder (**b**)

measurement was followed by an examination of wear mechanisms on the tool after 84 repeated plunging procedures, as shown in Fig. 9.

The partial view in Fig. 9a shows the probe at a magnification of 30:1. With a magnification of 110:1, material adherence was detected on the thread flank facing in the direction of the weld root. This relates to adherent aluminum caused by plunging and not removed during cleaning with NaOH. It can, therefore, be assumed that adhesive wear is already present during plunging. The detailed view of the shoulder in Fig. 9b exhibits isolated pitting at a magnification of 500:1. The clusters shown in Fig. 8 brought about a notch in the base material. The cyclic rotation of the FSW tool causes surface fatigue, leading to the removal of these clusters and the formation of pits. The clusters represent a mechanical notch in the base material from which crack growth can start. In comparison to the tool wear mechanisms after 84 plunging procedures, Fig. 10 depicts the wear mechanisms on shoulder and probe after 84 plunging procedures and a weld seam length of 80 m.

The analysis of the tool was carried out at the probe and shoulder. Closer observations were made for two areas at the probe. Figure 10a depicts the major thread radius. In addition to a completely worn surface, layer formation is visible at magnifications of 100:1 and 220:1. It can be assumed that a diffusion layer has formed between the aluminum and the tool material, which is caused by tribochemical reaction. In

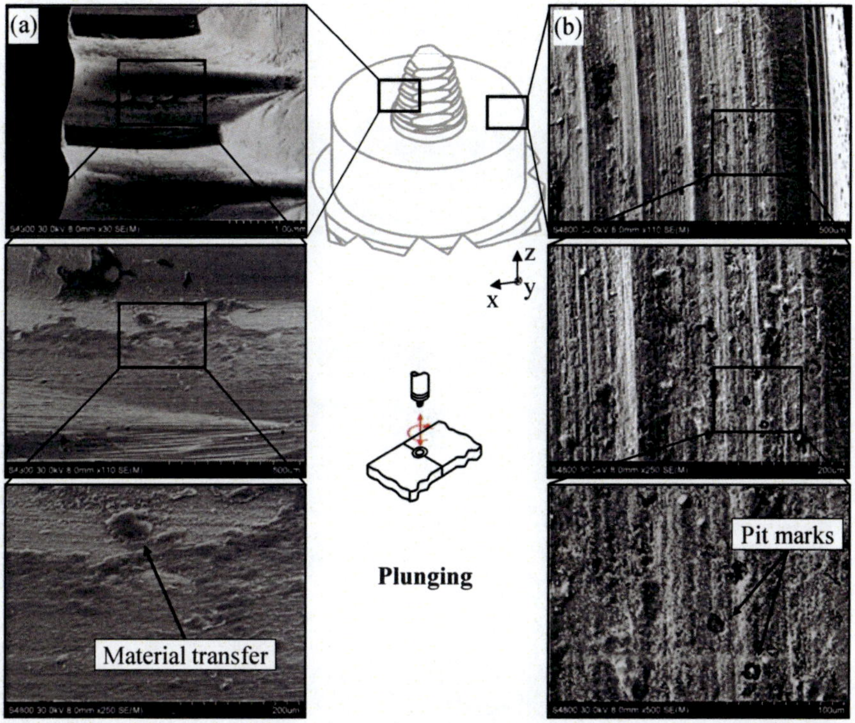

Fig. 9 SEM analysis of the FSW tool after 84 plunging procedures at the probe (**a**) and the shoulder (**b**). (Color figure online)

order to confirm this behavior, further investigations such as the determination of the crystal orientation and the evaluation of the layer composition are necessary. The layer formation is also visible on the shoulder surface in Fig. 10c; in both cases, the layer is interspersed with cracks. The cause of the cracks can be explained by the reduced ductility of the layer. Multi-axial stresses, caused by surface fatigue, cannot be relieved, leading to cracks. Reaching a critical thickness as well as through superimposed surface fatigue, the diffusion layer tends to break out brittle and additional abrasive wear particles are formed. Figure 10b shows the interaction between the penetrating abrasive and the probe. This indicates that the wear particles move along with the material flow, resulting in three-body abrasion. The results show that abrasion can occur during welding of soft aluminum alloys. Basically, it could be shown that FSW of AA-6060 T66 with tools made of H13 tool steel causes wear from a combination of adhesion, tribochemical reaction, surface disruption, and abrasion.

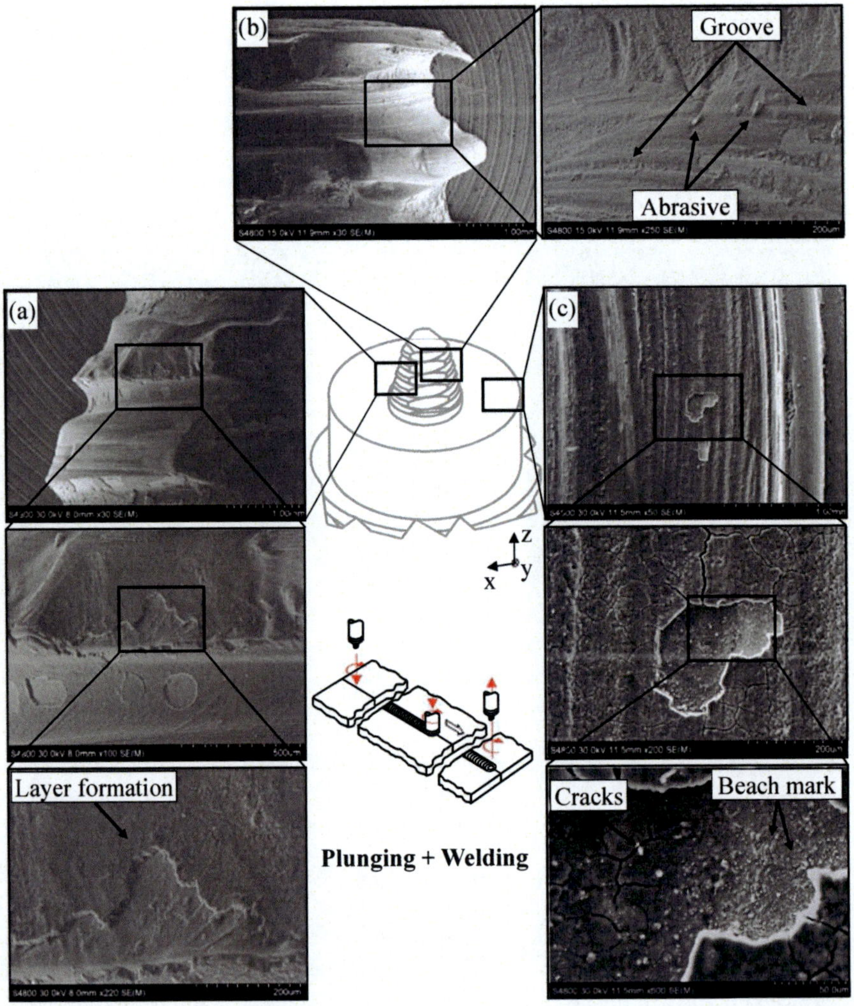

Fig. 10 SEM analysis of the FSW tool after 84 plunging procedures and a weld seam length of 80 m at the probe (**a**), (**b**) and the shoulder (**c**). (Color figure online)

Conclusions

In this study, the analysis of the effective wear mechanisms at shoulder and probe was investigated, disaggregating wear arising from plunging, and welding. The investigation were carried out using FSW tools made of H13 tool steel to join AA-6060 T66 aluminum alloy with a thickness of 8 mm. For the disaggregated analysis of tool wear, 21, 42, 63, and 84 plunging procedures were used to determine the corresponding weld seam lengths of 20, 40, 60, and 80 m. From the study above, the following conclusions can be drawn:

- Within 3D surface measurement, the disaggregation of plunging and welding shows that most of the FSW tool wear occurs during welding. This can be explained by the combined thermal and mechanical stresses. However, slight wear was observed during plunging, both on the side of the flank facing the direction of rotation and on the thread flanks facing in the direction of the weld root.
- The examination of wear mechanisms after 84 plunging procedures showed adherent aluminum on the thread flanks that faced in the direction of the weld root. Consequently, it can be assumed that adhesive wear is already present during plunging. Additionally, pit marks were identified on the shoulder surface due to the removal of carbides through surface fatigue.
- In contrast, the probe and shoulder exhibit a completely worn surface during welding. The formation of diffusion layers due to tribochemical reactions was thereby identified. In the case of critical layer thickness or superimposed surface fatigue, these tend to break out brittle due to their low ductility. Abrasive wear was identified, indicating that the wear particles move with the material flow, resulting in three-body abrasion. Consequently, it could be shown that the superimposed wear mechanisms lead to self-damaging of the FSW tool

Based on the wear mechanisms described for the shoulder and probe, it is possible to initiate demand-adapted improvements to increase tool life.

References

1. Ostermann F (2007) Anwendungstechnologie Aluminium. Berlin/Heidelberg, Germany, Springer Vieweg
2. Karakizis PN, Pantelis DI, Dragatogiannis DA, Bougiouri VD, Charitidis CA (2019) Study of friction stir butt welding between thin plates of AA5754 and mild steel for automotive applications. Int J Adv Manuf Technol 102:3065–3076
3. Goyal A, Garg RK (2019) Parametric optimization of friction stir welding process for marine grade aluminum alloy. Int J Struct Integr 10:162–175
4. Reddy SNJ, Sathiskumar R, Kumar KG, Jerome S, Jebaraj AV, Arivazhagan N Manikandan M (2019) Friction based joining process for high strength aerospace Aluminum alloy. Mater Res Express: 6
5. Schmidt HNB, Dickerson TL, Hattel JH (2006) Material flow in butt friction stir welds in AA2024-T3. Acta Materialia 54 Jg. Nr 4:1199–1209
6. Reynolds AP (2008) Flow visualization and simulation in FSW. Scripta materialia 58. Jg. Nr 5:338–342
7. DIN EN ISO 25239–1:2011 (2012) Rührreibschweißen – Aluminium – Teil 1 Begriffe
8. Voellner G (2009) Rührreibschweißen mit Schwerlast-Industrierobotern. Ph.D. Thesis, Technical University of Munich, Munich, Germany
9. Grätzel M, Regensburg A, Hasieber M, Gerken JA, Schürer R, Bergmann JP (2019) Scaling effects during friction stir welding of aluminum alloys with reduced tool aspect ratios. Weld World 63:337–347
10. Hasieber M, Grätzel M, Bergmann JP 2020 A novel approach for the detection of geometric-and weight-related FSW tool wear using stripe light projection. J Manuf Mater Process 4(2):60
11. Więckowski W, Burek R, Lacki P, Łogin W (2019) Analysis of wear of tools made of 1.2344 steel and MP159 alloy in the process of friction stir welding (FSW) of 7075 T6aluminum alloy sheet metal. EksploatacjaiNiezawodność 21:54–59

12. Thompson BT (2010) Tool Degradation Characterization in the Friction Stir Welding of Hard Metals, Graduate Program in Welding Engineering. The Ohio State University, Columbus OH USA
13. Eff M (2012) The Effects of Tool Texture on ToolWear in Friction StirWelding of X-70 Steel. Ph.D. Thesis, The Ohio State University: Columbus, OH, USA
14. Wang J, Su J, Mishra RS, Xu R, Baumann JA 2014 Tool wear mechanisms in friction stir welding of Ti–6Al–4V alloy. Wear 321:25–32
15. Fall A, Fesharaki MH, Khodabandeh AR, Jahazi M (2016) Tool wear characteristics and effect on microstructure in Ti-6Al-4V friction stir welded joints. Metals 6(11):275
16. Prado RA (2001) Friction StirWelding: A Study of ToolWear Variation in Aluminum Alloy 6061+ 20% Al2O3. Frict Stir Weld Process 2001:105–116
17. Prado RA, Murr LE, Soto KF, McClure JC (2003) Self-optimization in tool wear for friction-stir welding of Al 6061+ 20% Al2O3 MMC. Mater Sci Eng 349:156–165
18. Fernandez GJ, Murr LE (2004) Characterization of tool wear and weld optimization in the friction-stir welding of cast aluminum 359+ 20% SiC metal-matrix composite. Mater Charact 52:65–75
19. Shindo DJ, Rivera AR, Murr LE (2002) Shape optimization for tool wear in the friction-stir welding of cast AI359-20% SiC MMC. J Mater Sci 37:4999–5005
20. Weinberger T, Khosa SU, Führer B, Enzinger N (2008) Analysis of tool wear and failure mechanism during friction stir welding of steel. In: Proceedings of the 7th International Symposium Friction Stir Welding, Awaji, Japan, 20–22 May 2008
21. Sahlot P, Jha K, Dey GK, Arora A (2018) Wear-induced changes in FSW tool pin profile: Effect of process parameters. Metall Mater Trans A 49:2139–2150
22. Choi DH, Lee CY, Ahn BW, Choi JH, Yeon YM, Song K, Jung SB (2009) Frictional wear evaluation of WC–Co alloy tool in friction stir spot welding of low carbon steel plates. Int J Refract Metal Hard Mater 27(6):931–936
23. Czichos H, Habig KH (2010) Tribologie-Handbuch: Tribometrie, Tribomaterialien, Tribotechnik, 4th edn. Wiesbaden, Germany, Springer Vieweg
24. Zum KH (1985) Tribologie: Reibung-Verschleiß Schmierung. Naturwissenschaften 72:260–267
25. Hoßfeld M 2016 Experimentelle, Analytische und Numerische Untersuchungen des Rührreibschweißprozesses. Ph.D. Thesis, University of Stuttgart, Stuttgart, Germany
26. Tarasov SY, Rubtsov VE, Kolubaev EA (2014) A proposed diffusion-controlled wear mechanism of alloy steel friction stir welding (FSW) tools used on an aluminum alloy. Wear 318:130–134
27. Prater TJ, Strauss AM, Cook GE, Gibson BT, Cox CD (2013) A comparative evaluation of the wear resistance of various tool materials in friction stir welding of metal matrix composites. J Mater Eng Perform 22(6):1807–1813
28. Tiwari A, Pankaj P, Biswas P, Kore SD, Rao AG (2019) Tool performance evaluation of friction stir welded shipbuilding grade DH36 steel butt joints. Int J Adv Manuf Technol 103(5–8):1989–2005
29. Adesina AY, Al- FA, Gasem ZM (2018) Wear resistance performance of AlCrN and TiAlN coated H13 tools during friction stir welding of A2124/SiC composite. J Manuf Process 33:111–125

Friction Stir Lap Welding Between Al and FeCoCrNiMn High Entropy Alloy

Haining Yao, Ke Chen, Muyang Jiang, Min Wang, Xueming Hua, Lanting Zhang, and Aidang Shan

Abstract Friction stir lap welding between single-phase FeCoCrNiMn high entropy alloy (HEA) and commercial pure aluminum (1060Al) was studied. Sound joints were produced at a traverse speed of 50 mm/min and a rotation speed of 1500 RPM. The tensile shear strength reached ~240 N/mm. The joint fractured at Al base material, showing a high interfacial bonding. The microstructure of joints was discussed. The interfacial structures were composed of the composite structure in the Al side, layered structure in the HEA side, and a transition layer between these two structures. In addition, significant differences in the microstructure were observed from the 1060Al side to HEA side through the interfacial microstructure. The microstructure of the HEA exhibits a complex microstructure including an equiaxed structure and an elongated structure and the average grain size was measured from nanometer to micrometer. And aluminum grains close to the weld interface were equiaxed with a complete mixture of grain orientations.

Keywords High entropy alloy · Aluminum · Friction stir lap welding · Dissimilar joint

Introduction

High entropy alloys (HEAs) have attracted intensive attention in recent years [1]. Different from the traditional alloy design concepts, HEAs as novel multi-principal element alloys are not based on only one or two principal elements, but contain at least five major elements in equiatomic or near equiatomic ratio [2]. Owing to the unique multi-principal element composition, HEAs can possess special properties, including high strength and hardness, good structural stability, and excellent cryogenic properties [3, 4].

The studies of HEAs are mainly focused on the composition design and performance improvement [5–8], but few on the welding of HEAs [9–16]. When a novel

H. Yao · K. Chen (✉) · M. Jiang · M. Wang · X. Hua · L. Zhang · A. Shan
School of Materials Science and Engineering, Shanghai Jiao Tong University, 800 Dong Chuan Road, Shanghai 200240, People's Republic of China
e-mail: chenke83@sjtu.edu.cn

Y. Hovanski et al. (eds.), *Friction Stir Welding and Processing XI*, The Minerals, Metals & Materials Series, https://doi.org/10.1007/978-3-030-65265-4_4

alloy is used in structural applications, its welding is inevitable for most of time. Wu et al. [15] studied the weldability of FeCoCrNiMn by electron beam welding and the mechanical properties of the joint could be comparable to those of the base material at both room and cryogenic temperatures. Oliveira et al. [11] studied gas tungsten arc welding of the FeCoCrNiMn HEA and found that the joints have a lower strength and ductility than the base material.

However, it should be noted that previous researches on the welding of HEAs mainly focused on the welding of similar HEAs. There were very few studying dissimilar welding of HEAs to other traditional alloys [17, 18]. As a solid-state method, friction stir welding (FSW) has great potential for joining dissimilar materials, which offers lower peak temperatures than conventional fusion welding [19]. In this study, the dissimilar welding of single-phase FCC FeCoCrNiMn alloy and 1060Al was studied by using friction stir lap welding. The aim is to investigate the effect of FSW process on microstructure and mechanical properties between FeCoCrNiMn and 1060Al dissimilar joints.

Experimental

A commercial pure aluminum (1060Al) and a single-phase equiatomic FCC FeCoCrNiMn HEA were used in this study. FeCoCrNiMn ingot was fabricated by vacuum induction melting under pure Ar atmosphere. The purity of the alloying elements was above 99.9%. Aluminum plate and FeCoCrNiMn plate were machined into rectangular samples with the dimensions of 50 mm*85 mm*4 mm and 50 mm*85 mm*2 mm, respectively. The welding tool was made of tungsten–rhenium (W–Re) with a shoulder diameter of 15 mm, root diameter of 6 mm, pin diameter and length of 4 mm and 4.4 mm, respectively, and the welding tilt angle was 2.5°. A schematic diagram of the welding process by friction stir lap welding is shown in Fig. 1.

Two sets of welding parameters (low heat input and high heat input) were designed in the study, as shown in Table 1. The welding temperature was recorded by an infrared camera imager (*FLIR A615*). It is worth noting that the peak temperature obtained by the thermal camera imager is not the actual peak welding temperature and can be lower than the temperature near the tool. The cross sections of the friction stir lap welding joints were observed using an optical microscope (OM, *LEICA DM 4000*). A scanning electron microscope (SEM, *MIRA3 TESCAN*) equipped with an energy dispersive X-ray spectrometer (EDS) was utilized to investigate the interfacial microstructure. The microstructure and deformed grains nearby the interface were examined by electron backscatter diffraction (EBSD, *GAIA3 GMU Model 2016*). The specimens for tensile shear testing were cut to 60 mm long and 10 mm wide, perpendicular to the welding seam. Tensile-shear tests were conducted at a constant crosshead speed of 1 mm/min.

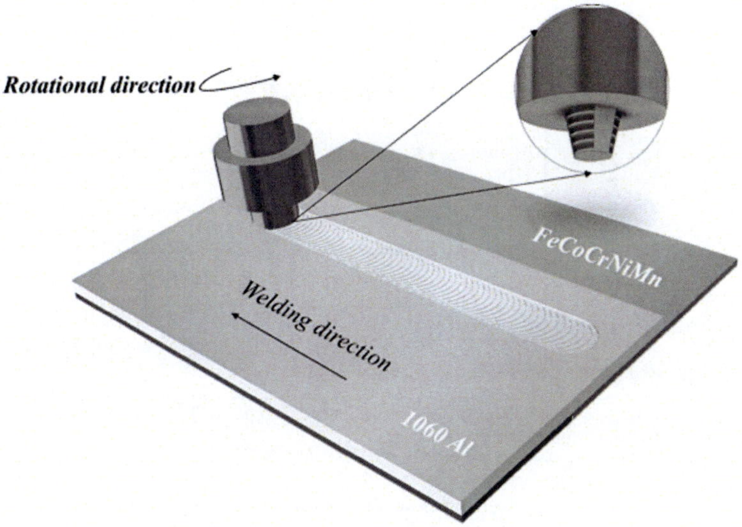

Fig. 1 Schematic of friction stir lap welding process and the configuration of tool used

Table 1 Welding parameters of different groups

Sample	Plunge depth (mm)	Plunge speed (mm·min^{-1})	Traveling speed (mm·min^{-1})	Rotational speed (rpm)
1	4.5	10	100	600
2	4.5	10	50	1500

Results and Discussion

Figure 2 shows the weld morphologies with low heat input (Sample 1) and high heat input (Sample 2). Burs were observed at the weld surface of Sample 1 due to the

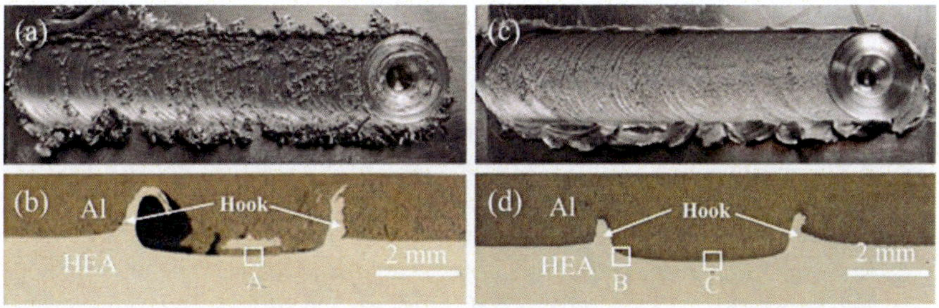

Fig. 2 Weld appearances and cross sections of FSW joints obtained at different parameters: **a**, **b** Sample 1 (600 RPM, 100 mm/min); **c**, **d** Sample 2 (1500 RPM, 50 mm/min). (Color figure online)

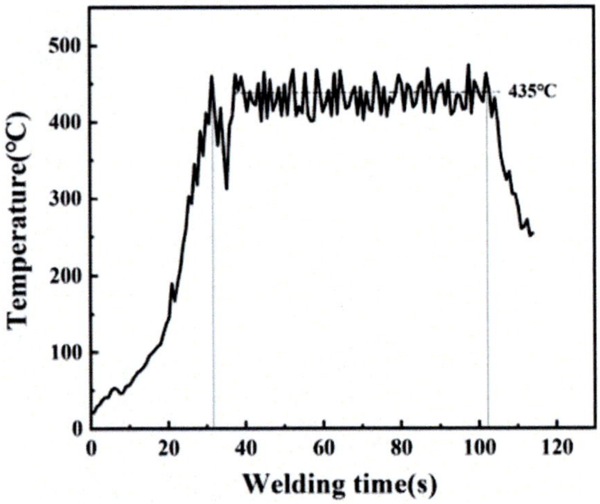

Fig. 3 Welding temperature variation along the weld seam surface of Sample 2

insufficient plastic flow, while the weld with high heat input is relatively smooth. Meanwhile, voids were found at the interfaces at the welding speeds of Sample 1, as shown in Fig. 2b, which has a significant effect on the tensile properties of the joint. It can be attributed to the insufficient heat input during the FSW. For Sample 2 of high heat input, no visible welding defect appeared in the joint. As the pin was inserted into FeCoCrNiMn at a depth of about 0.5 mm, the FeCoCrNiMn near the interface formed "Hook" structure under the direct effect of the pin, improving the mechanical interlocking between these two base materials.

During FSW, temperatures in the processed region were recorded by the infrared camera imager for real-time monitoring of welding process. As shown in Fig. 3, the welding temperature tended to be stable during welding, with the measured peak welding temperature relatively stable at around 435°C, which is about 2/3 of the melting point of aluminum alloy. This ensures good flowability of aluminum alloy in the welding process.

The welding interface between HEA and 1060Al was studied by SEM as shown in Fig. 4. For Sample 1, there was no obvious metallurgical bonding even at the center of the pin affected zone (Fig. 4a). For Sample 2, the interfacial structure between HEA and 1060Al can be divided into three different categories, i.e. the particles dispersion composite-like structure in the Al side (composite structure), layered structure in the HEA side and the transition layer between these two structures as shown in Fig. 4b and c. The particle size in the composite structure varied from nanometer to micrometer. According to the EDS result, the particle in the composite structure barely contained Cr and Mn. Al content was about 87.6%. Moreover, the phenomenon of complex-layered structure was characterized. Several swirl- and vortex-like layered structures could be observed, as shown in Fig. 4b. The formation of layered structure should be related to the flow pattern of material which was spiral movement within the rotational zone. It was noted that the thickness of the transition layer was non-uniform at the weld interface. A thicker transition layer was found at the interface marked C in

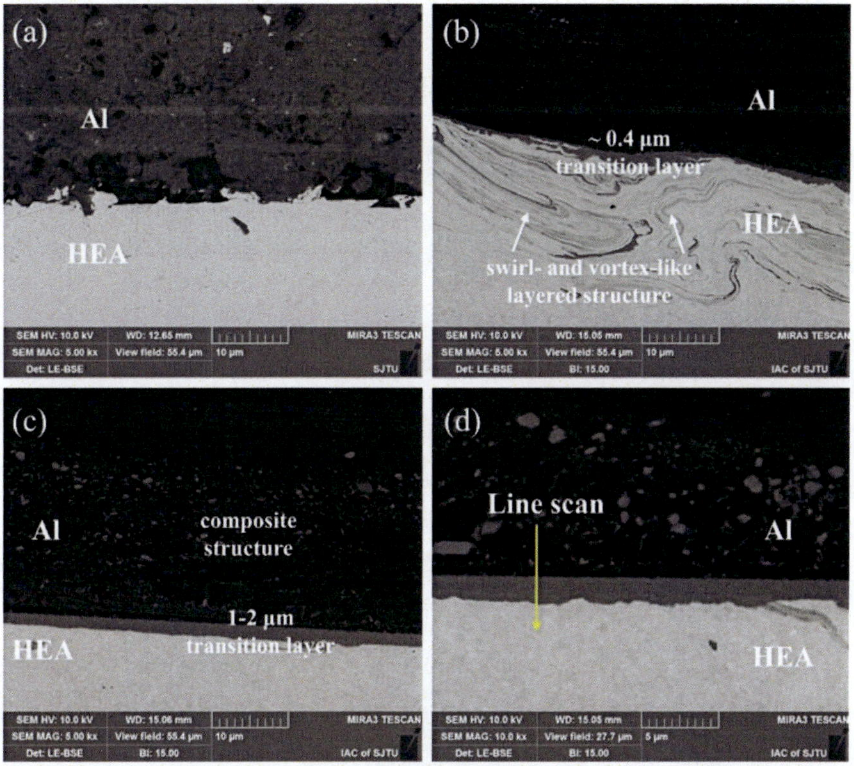

Fig. 4 SEM characterization marked in Fig. 2a: Magnified SEM view marked A in Fig. 2b; **b** Magnified SEM view marked B in Fig. 2d; **c** Magnified SEM view marked C in Fig. 2d; **d** Magnified SEM view at the interface of Sample 2. (Color figure online)

Fig. 2d. From the magnified view of Sample 2, the transition layer consisted of two discernible sublayers as shown in Fig. 4d. Figure 5 shows the result of the EDX line scan analysis. According to the variation of elements, it was deduced that the thickness of the transition interlayer at the interface was ~2 μm.

The interfacial microstructure for Sample 2 was further characterized by EBSD. Figure 6 shows the inverse pole figure (IPF) map. It can be seen that microstructure of the HEA exhibits a complex microstructure including an equiaxed structure and an elongated structure and the average grain size was measured from nanometer to micrometer. This could be attributed to compressive and shearing deformation applied by the rotating pin inserted into HEA. A partially recrystallized and gradient microstructure was thus formed. The above features of the HEA microstructure indicate that a fully dynamic recrystallization did not occur during the FSW process. Furthermore, no Kikuchi pattern was obtained in the 1-2 μm thick transition layer.

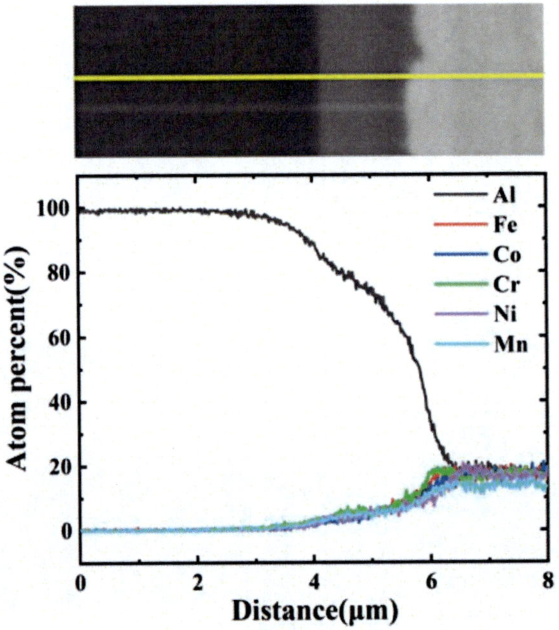

Fig. 5 SEM-EDS scanning line result marked in Fig. 4. (Color figure online)

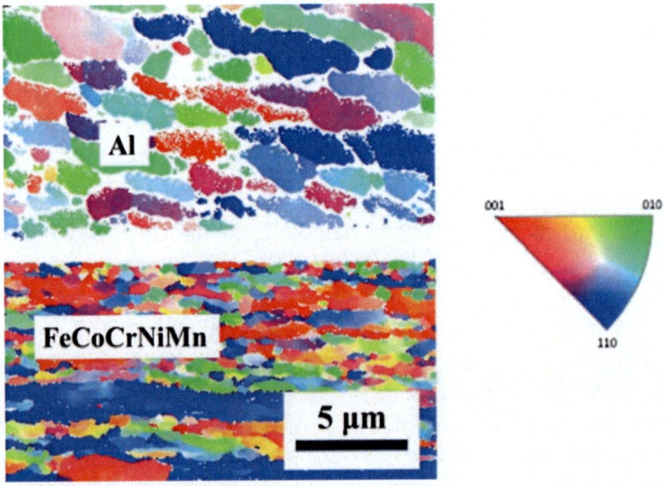

Fig. 6 Inverse pole figures obtained from regions corresponding to Fig. 4d. (Color figure online)

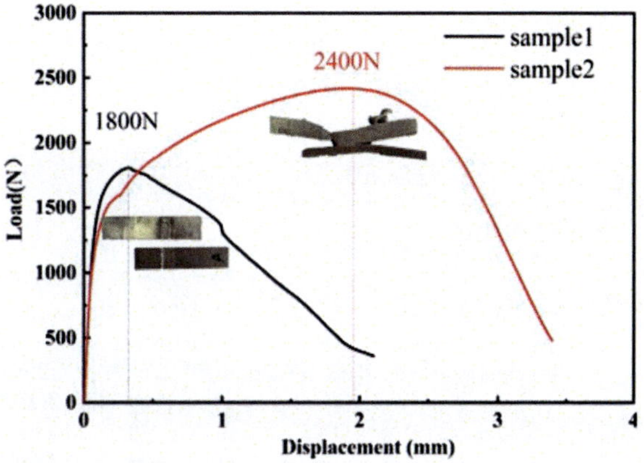

Fig. 7 Tensile shear strength of FSWed Al/FeCoCrNiMn joints. (Color figure online)

Aluminum grains close to the weld interface were equiaxed. This resulted from recrystallization due to shear deformation and deformation heating during FSW. The low-indexed rate in local region could be attributed to the composite structure. In addition, a complete mixture of grain orientations was observed in Fig. 6, in which the color variations of the grains qualitatively represent the different grain misorientations.

Figure 7 shows the results of the joint tensile and shear properties. Sample 1 cracked along the interface with strength of 1800 N. However, Sample 2 broke at the base material of 1060Al alloy and its strength reached 2400 N. The tensile test results showed that the bonding between HEA and 1060Al at the interface had an important effect on the joint strength. This indicated that a better FSW parameter could realize the high strength joining between HEA and 1060Al alloy.

Conclusions

Friction stir lap welding between single-phase FeCoCrNiMn HEA and commercial pure aluminum (1060Al) was studied. The as-welded joint microstructures and properties were investigated. The conclusions could be drawn as the following:

(1) Sound joints were produced at a traverse speed of 50 mm/min and a rotation speed of 1500 RPM, and the ultimate fracture occurred in Al base material.
(2) A metallurgical bonding was achieved at the interface. The interfacial structures are composed of the composite structure in the Al side, layered structure in the HEA side and a transition layer sitting in between.
(3) The microstructure of the HEA exhibits a complex microstructure including an equiaxed structure and an elongated structure and the average grain size was measured from nanometer to micrometer. And aluminum grains close to the weld interface were equiaxed with a complete mixture of grain orientations.

Acknowledgements This work was supported by the Natural Science Foundation of China (No. 52075330), Science and Technology Major Project of Yunnan Province (No. 2018ZE005-6) and the Shanghai Key Laboratory of Material Laser Processing and Modification (MLPM2015-2).

References

1. Yeh JW, Chen SK, Lin SJ, Gan JY et al (2004) Nanostructured high-entropy alloys with multiple principal elements: novel alloy design concepts and outcomes. Adv Eng Mater 6(5):299–303
2. Sathiyamoorthi P, Kim HS (2020)High-entropy alloys with heterogeneous microstructure: Processing and mechanical properties. Prog Mater Sci:100709
3. Li Z, Zhao S, Ritchie RO, Meyers MA (2019) Mechanical properties of high-entropy alloys with emphasis on face-centered cubic alloys. Prog Mater Sci 102:296–345
4. Zhang Y, Zuo TT, Tang Z, Gao MC et al (2014) Microstructures and properties of high-entropy alloys. Prog Mater Sci 61:1–93
5. Deng Y, Tasan CC, Pradeep KG, Springer H et al (2015) Design of a twinning-induced plasticity high entropy alloy. Acta Mater 94:124–133
6. Gao X, Lu Y, Zhang B, Liang N et al (2017) Microstructural origins of high strength and high ductility in an AlCoCrFeNi2.1 eutectic high-entropy alloy. Acta Mater 141:59–66
7. Liang YJ, Wang L, Wen Y, Cheng B et al (2018) High-content ductile coherent nanoprecipitates achieve ultrastrong high-entropy alloys. Nat Commun 9(1):1–8
8. Schuh B, Mendez-Martin F, Völker B, George EP et al (2015) Mechanical properties, microstructure and thermal stability of a nanocrystalline CoCrFeMnNi high-entropy alloy after severe plastic deformation. Acta Mater 96:258–268
9. Chen Z, Wang B, Duan B, Zhang X (2020)Mechanical properties and microstructure of laser welded FeCoNiCrMn high-entropy alloy. Mater Lett: 127060
10. Li P, Sun H, Wang S, Hao X et al (2020) Rotary friction welding of AlCoCrFeNi2.1 eutectic high entropy alloy. J Alloy Compd 814:152322
11. Oliveira JP, Curado TM, Zeng Z, Lopes JG et al (2020) Gas tungsten arc welding of as-rolled CrMnFeCoNi high entropy alloy. Mater Des 189:108505
12. Park S, Nam H, Park J, Na Y et al (2020) Superior-tensile property of CoCrFeMnNi alloys achieved using friction-stir welding for cryogenic applications. Mater Sci Eng A: 139547
13. Qin X, Xu Y, Sun Y, Fujii H et al (2020) Effect of process parameters on microstructure and mechanical properties of friction stir welded CoCrFeNi high entropy alloy. Mater Sci Eng A: 139277
14. Sokkalingam R, Sivaprasad K, Duraiselvam M, Muthupandi et al (2020) Novel welding of Al0.5CoCrFeNi high-entropy alloy: corrosion behavior. J Alloy Compd:817:153163
15. Wu Z, David SA, Feng Z, Bei H (2016) Weldability of a high entropy CrMnFeCoNi alloy. Scripta Mater 124:81–85
16. Yang X, Dong P, Yan Z, Cheng B et al (2020) AlCoCrFeNi high-entropy alloy particle reinforced 5083Al matrix composites with fine grain structure fabricated by submerged friction stir processing. J Alloy Compd: 155411
17. Nene SS, Gupta S, Morphew C, Mishra RS (2020) Friction stir butt welding of a high strength Al-7050 alloy with a metastable transformative high entropy alloy. Materialia: 100740
18. Wang G, Sheng G, Sun J, Wei Y et al (2020) Mechanical properties and microstructure evolution of CrMnFeCoNi HEA/304 SS dissimilar brazing joints. J Alloy Compd: 154520
19. Mishra RS, Ma ZY (2005) Friction stir welding and processing. Mater Sci Eng R Rep 50(1–2):1–78

Modified Friction Stir Welding of Al–Zn–Mg–Cu Aluminum Alloy

Ahmad Alali Alkhalaf, Anna Tesleva, Pavel Polyakov, Matthias Moschinger, Sebastian Fritsche, Iuliia Morozova, Anton Naumov, Fedor Isupov, Gonçalo Pina Cipriano, and Sergio T. Amancio-Filho

Abstract Friction Stir Welding (FSW) is adopted to join the Al-Zn-Mg-Cu (7xxx) aluminum alloys in order to improve strength and ductility of the joints. The paper presents an investigation on the effect of the modified FSW methods such as Impulse FSW (IFSW) and High-Speed FSW (HS-FSW) on the microstructure and mechanical properties of the AA7075 T6 aluminum alloy. It was demonstrated that the tensile properties and hardness of the IFSW joints could be improved compared to the traditional FSW joints by variation of the impulse parameters (force and frequency). This enhancement is caused by microstructural changes. The brittle fracture of the HS-FSW joints is explained by the presence of defects in the weld zone. It indicates that heat input was insufficient or intermixing in the stir zone was inadequate.

Keywords Al–Zn–Mg–Cu aluminium alloy · Friction stir welding · Impulse friction stir welding · High-speed friction stir welding

Introduction

Al-Zn-Mg-Cu series (7xxx) aluminium alloys are used in the aerospace, automotive, marine, and rail industries due to their outstanding properties such as high strength and toughness, low density, and outstanding machinability [1, 2]. Among disadvantages of 7xxx series alloys, low ductility, increased sensitivity to stress concentration, high anisotropy of properties, and tendency to stress corrosion cracking could be highlighted.

The main alloying elements of 7xxx series aluminium alloys are zinc, magnesium, and copper which are responsible for precipitation strengthening mechanism

---

A. Alali Alkhalaf (✉) · A. Tesleva · P. Polyakov · I. Morozova · A. Naumov · F. Isupov
Peter the Great St. Petersburg Polytechnic University, St. Petersburg, Russia
e-mail: ahmadakelms@gmail.com

I. Morozova
Brandenburg University of Technology Cottbus-Senftenberg, Brandenburg, Germany

M. Moschinger · S. Fritsche · G. Pina Cipriano · S. T. Amancio-Filho
Graz University of Technology, Graz, Austria

Y. Hovanski et al. (eds.), *Friction Stir Welding and Processing XI*, The Minerals, Metals & Materials Series, https://doi.org/10.1007/978-3-030-65265-4_5

through the formation of various precipitates such as $MgZn_2$, Al_2Cu, Al_2CuMg, and $Mg_4Zn_3Al_3$ [3–5]. Manganese and chromium enhance ageing effect and increase the corrosion resistance of the alloy. In addition, manganese promotes the formation of a fine-grained structure, prevents the precipitation of intermetallic phases along the grain boundaries of the solid solution and increases the strength of the alloy.

One of the significant challenges in joining Al-Zn-Mg-Cu alloys using conventional fusion techniques is that it leads to formation of various defects in the weld and deterioration of the strength and ductility of the joint. Better joint properties can be achieved when Friction Stir Welding (FSW) is adopted as the defects that may result from fusion welding could be prevented [6]. However, the strengthening precipitations can change during FSW process due to their heat-sensitivity which results in joint softening [7–11]. As a result of the transformations in the precipitation sequence, a typical "W" shape of the microhardness profile appears with the hardness decrease in the heat-affected zone (HAZ) of the weld. In order to avoid this problem, different measures are taken such as underwater FSW or High-Speed FSW in precipitation-hardenable aluminium alloys [5, 11–15]. It is reported that an increase in the welding speed led to a narrowing of the HAZ and hardness improvement in this zone, which resulted in a positive effect on the weld yield strength efficiency [14, 15].

In the current work, the study and comparison of the welds produced with standard friction stir welding, high-speed FSW and impulse FSW are presented. The macrostructures and mechanical properties of the welds were investigated in order to evaluate the feasibility of the modified FSW techniques to AA7075 T6 aluminum alloy and influence of the specific welding parameters on the joint quality.

Experimental Procedure

The chemical composition of the AA7075 T6 (AlZnMgCu1,5) precipitation-hardenable aluminum alloy used in the study is shown in Table 1.

Standard FSW, impulse FSW, and high-speed FSW were performed parallel to the rolling direction of the 1,7 mm thick sheets by means of a five-axis FSW machine Matec 40P with different welding parameters. The best parameters set for each type of FSW techniques were selected according to the investigation of the weld surface and are presented in Table 2.

The tool used for the performing of all FSW techniques was composed of a tapered tool probe with a diameter of 4 mm on the top and 3 mm on the bottom as well as a length of 1,6 mm. The diameter of a smooth concave shoulder was 10 mm.

Table 1 Chemical composition of the AA7075 alloy (wt.%)

Fe	Si	Cu	Mn	Mg	Cr	Zn	Al
0.19	0.65	1.4	0.16	2.92	0.28	6.70	Bal

Table 2 Welding parameters of the applied FSW techniques

FSW technique	Rotation speed [rpm]	Weld speed [mm/min]	Force [kN]	Frequency [Hz]
Standard FSW	1000	300	3,2	–
Impulse FSW	1000	300	3,2 ± 2	2
High-Speed FSW	2000	2000	6,5	–

The principle of IFSW was described in the previous work [16–19]. In comparison to FSW, axial force during the welding stage of IFSW is variable and results in time-dependent force, torque, contact condition at the tool/workpiece interface and heat input. The main advantage of IFSW is the improvement of mechanical properties of the welds. The addition of the impulses has a positive effect on the increase in the fatigue strength and ductility [18, 19]. A previous study of IFSW butt welds made of AA 6082-T6 alloy [18] has confirmed an increase of the fatigue strength close to 40%, compared to conventional FSW. The IFSW was realized on aluminum alloy 7075 T6. According to the obtained results from the previous work [16–19], there was no influence of the additional impulses of high frequency and force (higher than 2 Hz) on the mechanical properties of the produced joint. Therefore, IFSW by the frequency of 2 Hz and the force of 2 kN was performed in the current study. A sinusoidal function, as displayed in Fig. 1, was used as an input signal to change the applied force between 1,2 kN and 5,2 kN.

The samples for optical microscopy were cut using a slow speed diamond cutter under water-cooling and cold mounted in acrylic resins in order to avoid a thermal influence on the microstructure. After that the samples were mechanically ground from 300 to 2000 grade SiC paper, polished to 3 μm diamond suspensions, followed by a final polishing stage using a 0.05 μm colloidal-silica suspension. Etching of the polished specimens was carried out using Keller reagent (95 ml water; 2,5 ml HNO_3; 1,5 ml HCl; 1 ml HF) for 1 min for the optical investigation with polarization contrast. The Vickers microhardness was measured along the cross section of the weld with a load of 0.980 N (HV 0.1) for 10 s.

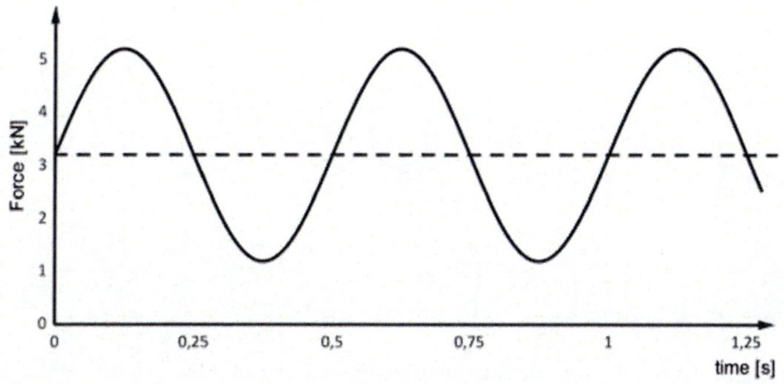

Fig. 1 Sinusoidal force signal for impulse welding with a frequency of 2 Hz

Tensile test was carried out on a tensile machine Zwick/Roell Z100 using 10 MPa/s loading speed at room temperature. The trials were performed up to the rupture of the tested specimens. The tensile strength of the weld was determined by testing five specimens of each weld.

Results and Discussion

Analysis of Defects

It is well known that the cross section of the FSW joint can be divided into different zones: Stir zone (SZ), thermomechanical affected zone (TMAZ), heat-affected zone (HAZ), and base material (BM), as demonstrated in Fig. 2. All produced joints were investigated by means of optical microscopy regarding the defects to analyze the influence of various parameters on the weld quality.

The aluminum oxide (Al_2O_3), which forms on the surface of the aluminum sheets, doesn't dissolve during the FSW process as it has a high melting point, therefore it remains in the SZ as a result of stirring action during FSW in the form of the remnant oxide lines (ROL). The ROL are considered to have a negative impact on the strength characteristics of the FSW joint. Figure 3 shows the ROL (depicted with the white arrows) that occurred in the SZ of the FSW and IFSW joints. Otherwise, no defects were revealed in the mentioned welds.

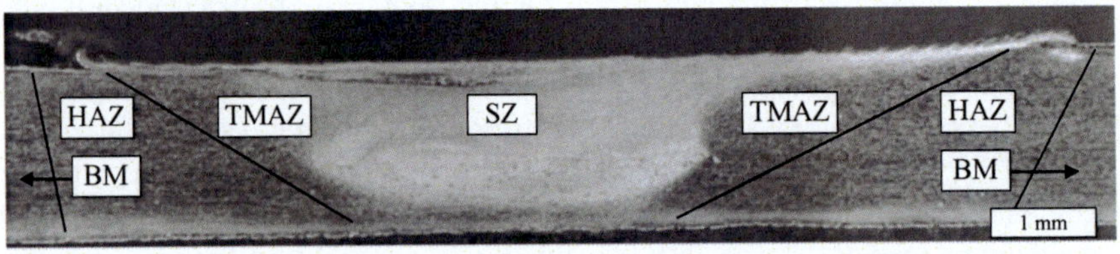

Fig. 2 Weld zones in the standard FSW joint

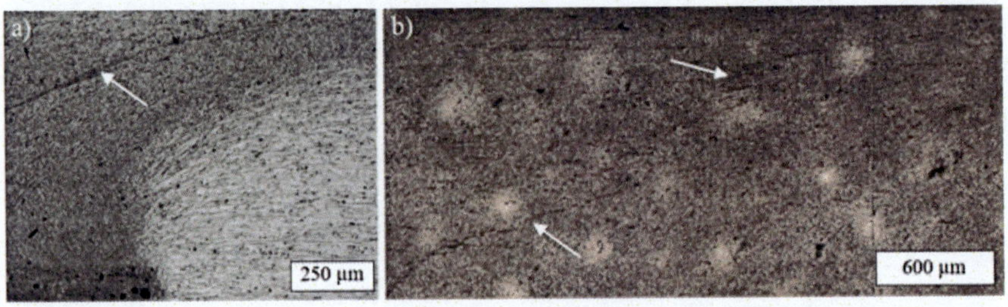

Fig. 3 ROL in the stirring zone after **a** Standard FSW; **b** Impulse friction stir welding. (Color figure online)

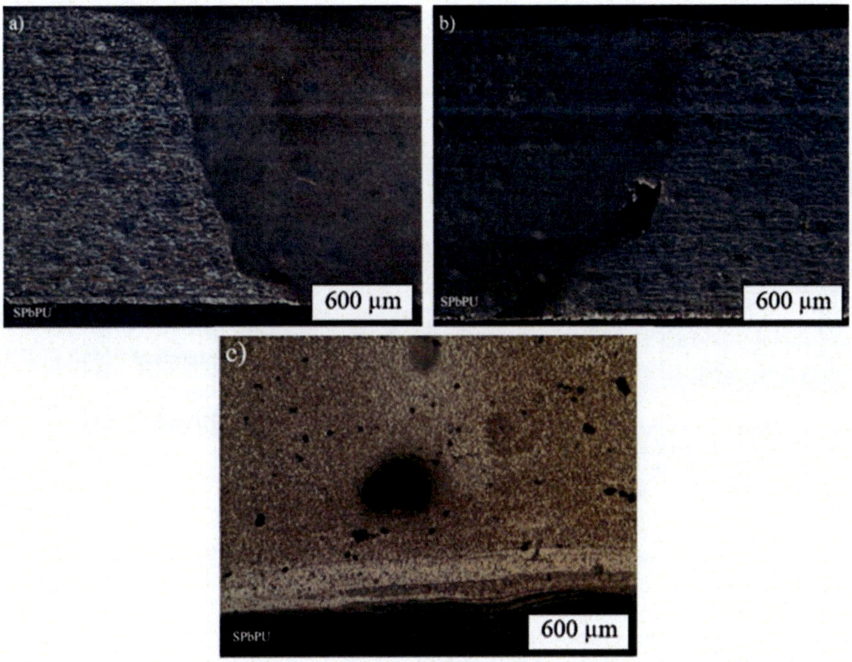

Fig. 4 Tunnel defect at **a** advancing side; **b** retreating side; **c** the wormhole in high-speed FSW joint. (Color figure online)

The tunnel defect was observed in the root area at both advancing and retreating side of the HS-FSW weld, as shown in Fig. 4a, b. This defect can form because of insufficient mixing due to low processing temperature [18–20]. In order to avoid the formation of the tunnel defect, it is recommended to balance the welding and rotation speed. Change of a tool configuration to the tapered probe with thread on the shoulder could also improve stirring action. In addition, there was a wormhole in the weld zone, as demonstrated in Fig. 4c. A possible reason for this defect formation could be excessive heat input during welding [19].

Microhardness Measurement

Figure 5 shows the results of the microhardness measurement through the thickness centerline for the cross sections of all produced welds. By comparing the three FSW methods, it can be concluded that the standard FSW and IFSW joints possessed similar hardness distribution across the weld section. It can be characterized by a typical hardness decrease in the transition region TMAZ/HAZ; however, it can be highlighted that the hardness in both HAZ and SZ was improved by IFSW process. A pronounced hardness increase in the HAZ region as well as its narrowing when

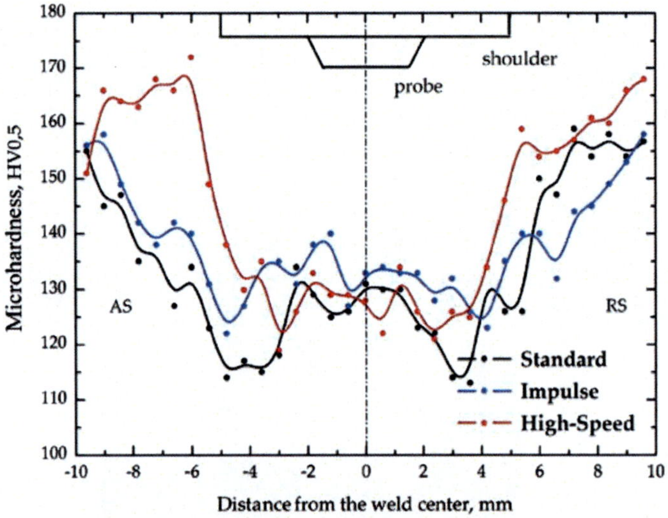

Fig. 5 Microhardness of Standard FSW; IFSW and HS-FSW joints. (Color figure online)

compared to the conventional FSW joint can be detected in case of HS-FSW. Taken together, the lowest hardness in the TMAZ was observed in the standard FSW joint, demonstrating the positive effect of the modified methods.

Tensile Test

The tensile strength of the produced FSW joints is presented in Fig. 6. The results indicate that the tensile strength of the welds is lower than that of the base material, which has a minimum value of 510 MPa. The highest tensile strength was achieved with IFSW. The welds produced by high-speed FSW possessed the lowest value.

Figures 7, 8 and 9 show the fracture locations of the welded joints. The fracture of the standard FSW and IFSW joints occurred in the TMAZ, where the lowest hardness value was indicated, as presented in Figs. 7 and 8. In contrast, the brittle fracture in the TMAZ or inside the SZ took place in case of HS-FSW joints (Fig. 9) which was a reason for the lowest tensile strength. Brittle fracture could be explained by the presence of a tunnel defect in the root zone, which is a strong stress concentrator.

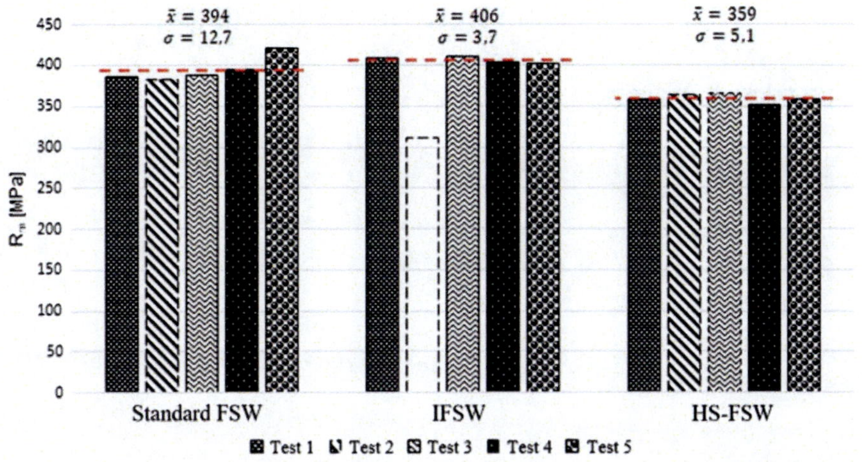

Fig. 6 Tensile strength of the produced welds. (Color figure online)

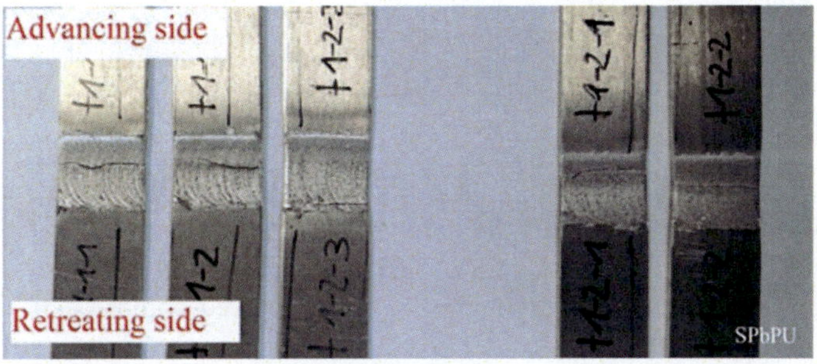

Fig. 7 The fracture perpendicular to the tensile direction of the standard FSW specimens. (Color figure online)

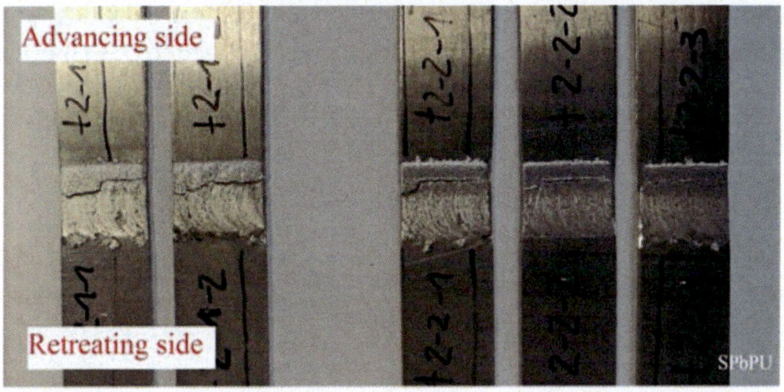

Fig. 8 The fracture perpendicular to the tensile direction of the IFSW specimens. (Color figure online)

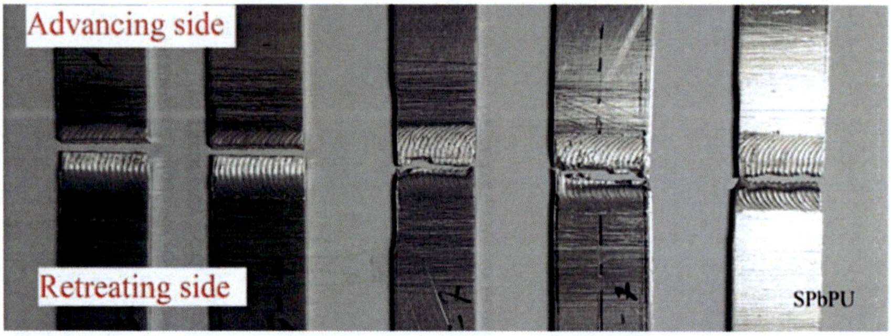

Fig. 9 The brittle fracture perpendicular to the tensile direction of top side of the High-Speed FSW. (Color figure online)

Conclusion

AA7075 T6 aluminum alloy was subjected to the three FSW methods: impulse FSW, high-speed FSW, and standard FSW as a reference technique in order to determine the influence of the welding parameters on the joint properties. The following results were obtained:

1. Standard FSW and IFSW joints are similar in terms of the hardness distribution across the weld; however, the hardness in the transition region was enhanced by IFSW compared to the FSW. Using HS-FSW, the HAZ was significantly narrowed, and thereby the hardness was improved compared to the standard friction stir weld.
2. The tensile strength of the welds is lower than that of the base material. The highest tensile strength is achieved with IFSW. The welds produced by high-speed FSW possessed the lowest UTS value, due to the presence of defects. It is necessary to optimize tool geometry to obtain defect free welds in future work.
3. The fracture of the standard FSW and IFSW joints occurred in the TMAZ, where the hardness value was the lowest. In contrast, the fracture of the HS-FSW joints was brittle due to the presence of a tunnel defect in the root zone, which is a strong stress concentrator.

Acknowledgements This research work was supported by the Academic Excellence Project 5–100 proposed by Peter the Great St. Petersburg Polytechnic University.

References

1. Sharma MM, Amateau MF, Eden TJ (2005) Mesoscopic structure control of spray formed high strength Al–Zn–Mg–Cu alloys. Acta Mater 53:2919–2924
2. Li LI, Tietao Z, Huan-xi L (2006) Effect of additional elements on aging behavior of Al–Zn–Mg–Cu alloys by spray forming. Trans Nonferr Metal Soc China 16:532–538
3. Sha G, Wang YB, Liao XZ, Duan ZC, Ringer SP, Langdon (2009) Influence of equal channel angular pressing on precipitation in an Al–Zn–Mg–Cu alloy. Acta Mater 57:3123–3132
4. Liu Y, Jiang D, Li B (2014) Effect of cooling aging on microstructure and mechanical properties of an Al–Zn–Mg–Cu alloy. Mat Des 57:79–86
5. Wang Q, Zhao Z, Zhao Y (2016) The strengthening mechanism of spray forming Al-Zn-Mg-Cu alloy by underwater friction stir welding. Mat Des 102:91–99
6. Langari J (2016) Effect of tool speed on axial force, mechanical properties and weld morphology of friction stir welded joints of A7075-T651. Int J Eng (2016):403–410
7. Barcellona A, Buffa G, Fratini L (2006) On microstructural phenomena occurring in friction stir welding of aluminum alloys. J Mat Proc Tech 177:340–343
8. Fu R, Sun Z, Sun R (2011) Improvement of weld temperature distribution and mechanical properties of 7050 aluminum alloy butt joints by submerged friction stir welding. Mat Des 32:4825–4831
9. Venugopal T, Srinivasa Rao K, Prasad Rao K (2004) Studies on friction stir welded aa 7075 aluminum alloy. Trans Indian Inst Metal 57:659–663
10. Hassan KhAA, Prangnell PB, Norman AF (2003) Effect of welding parameters on nugget zone microstructure and properties in high strength aluminium alloy friction stir welds. Sci Technol Weld Join 8(4):257–268
11. Upadhyay P, Reynolds AP (2010) Effects of thermal boundary conditions in friction stir welded AA7050-T7 sheets. Mater Sci Eng A 527:1537–1543
12. Wang Q, Zhao Z (2015) The adjustment strategy of welding parameters for spray formed 7055 aluminum alloy underwater friction stir welding joint. Mat Des 88:1366–1376
13. Zhao Y, Wang Q (2014) Microstructure and mechanical properties of spray formed 7055 aluminum alloy by underwater friction stir welding. Mat Des 56:725–730
14. Rodrigues DM, Leitão C, Louro R, Gouveia H, Loureiro H (2010) High speed friction stir welding of aluminium alloys. Sci Tech Weld Join 15:676–681
15. Liu H, Liu X, Wang X (2017) Mechanical properties and their relations to microstructural characteristics of high-speed friction stir-welded AA6005A-T6 aluminum hollow extrusions. Int J Adv Manuf Technol 88:3139–3149
16. Golubev I, Morozova I, Naumov A, Hantelmann C, Doynov N, Michailov V (2017) Numerical simulation and experimental investigation on impulse friction stir welding of 6082-T6 aluminum alloy. MS&T17: 987–994
17. Kondrat'ev SY, Morozova YN, Golubev YA, Hantelmann C, Naumov AA, Mikhailov VG (2018) Microstructure and mechanical properties of welds of Al–Mg–Si alloys after different modes of impulse friction stir welding. Metal Sci Heat Treat 59(11–12):697–702
18. Michailov V, Hantelmann C, Kloshek A (2011) Im-pulsrührreibschweißen – Ein Verfahren mit neuen Möglichkeiten. Große Schweißtechnische Tagung DVS 275:171–176
19. Morozova I (2016) Untersuchung des Einflusses der Impulskraft auf die Gefüge und die Eigenschaften der Impulsrührreibschweißverbindung. M. Sc. thesis, BTU Cottbus-Senftenberg, 112
20. Salih OS, Ou H, Sun W (2015) A review of friction stir welding of aluminium matrix composites. Mat Des 86:61–71

Part II
High Melting Temperature Materials

Low-Force Friction Surfacing for Crack Repair in 304L Stainless Steel

Hemant Agiwal, Hwasung Yeom, Kumar Sridharan, Kenneth A. Ross, and Frank E. Pfefferkorn

Abstract The objective of this research is to evaluate low-force friction surfacing as a means for repairing cracks in 304L stainless steel canisters. The motivation for reducing process forces is to allow for portability and miniaturization of the system to perform in-field repairs in confined spaces. 304L austenitic stainless steel rod was deposited over a substrate of the same alloy using a CNC machine tool in position control mode while observing the process forces on the back side of the repair. Friction surfacing was performed at spindle speeds up to 20,000 rpm. A fine grain structure was observed in the plastically deformed material. Axial pressures measured during the process showed a significant reduction when tests were performed at 20,000 rpm as compared to lower speeds. Friction surfacing was performed on simulated cracks followed by helium leak testing to evaluate the ability to create an acceptable repair. Closing of cracks near the interface of the substrate and deposit was observed after microstructural analyses. Acceptable helium leak rates through the repair can be achieved.

Keywords Friction surfacing · Stainless steel · Crack repair · Leak testing

H. Agiwal · F. E. Pfefferkorn (✉)
Department of Mechanical Engineering, University of Wisconsin-Madison, Wisconsin 53706, USA
e-mail: frank.pfefferkorn@wisc.edu

H. Yeom
Department of Engineering Physics, University of Wisconsin-Madison, Wisconsin 53706, USA

K. Sridharan
Departments of Engineering Physics and Materials Science & Engineering, University of Wisconsin-Madison, Wisconsin 53706, USA

K. A. Ross
Pacific Northwest National Laboratory, Richland, Washington 99354, USA

Y. Hovanski et al. (eds.), *Friction Stir Welding and Processing XI*, The Minerals, Metals & Materials Series, https://doi.org/10.1007/978-3-030-65265-4_6

Introduction

Friction surfacing is a solid-state material deposition process that produces fine-grained coatings with superior surface and corrosion properties [1]. Friction surfacing was first mentioned in a 1941 patent by Klopstock [2]; however, there has been a growing in the process over the last two decades. A schematic of friction surfacing is shown in Fig. 1a. A rotating rod is pressed against the substrate under an applied axial load. Frictional heat generates a viscoplastic boundary layer at the rod tip. The pressure and temperature conditions lead to an interdiffusion process resulting in a metallic bond between the plasticized material and the substrate. As seen in Fig. 1b, heat conduction into the substrate enables this layer to consolidate near the bonded interface, and as such, the viscoplastic shearing interface is formed between the rotating consumable rod and the top of the deposited layer. As the substrate translates past the face of the rotating tool, the material at the rubbing interface will either go towards developing the flash or will form the coating. With the on-going heat conduction, this viscoplastic shearing interface moves away from the substrate surface, increasing the thickness of the layer. By applying a traversing movement, the viscoplastic material is deposited onto the substrate surface in a continuous process. The difference in tangential speed of the consumable rod and traversing speed (V_x in Fig. 1a) of the substrate was also reported as the cause for deposits to detach [3]. This difference in speed also resulted in a difference in the flow patterns of marker material in the bottom and top half of the coatings as observed by Rafi et al. [4] by using tungsten marker material while depositing 304 stainless steel over mild steel substrates.

Considerable research has been conducted on friction surfacing of austenitic stainless steel. One of the first studies reported the aerial coverage applications by friction surfacing of 304 and 316 austenitic stainless steels on mild steel substrates [5]. The authors studied the impact of providing inclination to the consumable rod during the process. Higher bond strength was reported with an increasing inclination angle from 0° to –15°. Owing to the dynamic nature of phase transformation in stainless steel, a lot of emphases have been given to microstructural changes and material flow occurring during the process. Rafi et al. [6] reported a reduction in average

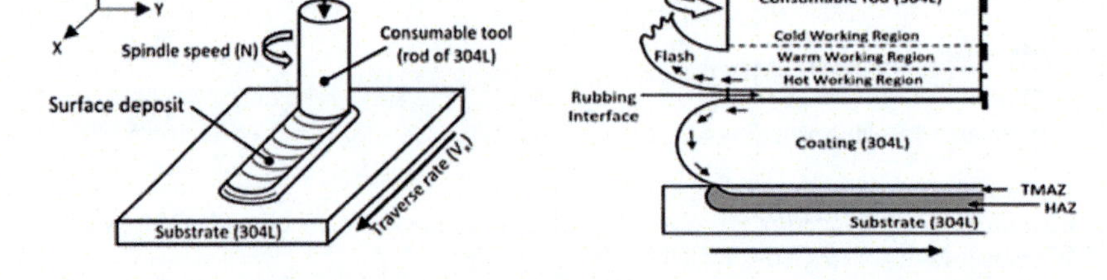

Fig. 1 **a** Schematic of friction surfacing and **b** thermomechanical events occurring during friction surfacing

grain size from 40 to 5 μm due to dynamic recrystallization in the deposited layer. Reduction in grain size was also observed by Puli and Ram [7], who reported that the average grain size increased from 4.8 μm at the coating-substrate interface to 9.4 μm near the top of the coating. They attributed this change to the reduction in the cooling rate at the top of the coating. The authors also found necklacing of sub-grains around existing grains, which are a classic sign of discontinuous dynamic recrystallization. Discontinuous dynamic recrystallization was also observed by Guo et al. [8]; however, they reported the grains to be in competing domination between discontinuous and continuous dynamic recrystallization as spindle speed is changed. Changes in strain rate and temperatures during the process with changing spindle speeds were given as the reason.

Austenitic stainless steels have been widely used across various industries such as marine, nuclear, and transport for its high corrosion resistance. The grade 304L which is a low carbon grade (<0.03%) has been extensively used in the nuclear industry to make the pressure vessels for storing used nuclear fuel. Irradiation leads to n/α reactions with boron and nickel present in the stainless steels and forms helium [9]. Even though the low C content in austenitic stainless steels helps in the reduction of sensitization, the chromium-rich alloy leads to the formation of sigma and chi phases at prolonged exposure to elevated temperatures. These sigma phases reduce the corrosion resistance of 304L stainless steel [10]. Helium embrittlement attacks these grain boundaries rich in sigma phases and makes them susceptible to viscous effects such as pitting and stress corrosion cracking (SCC). Various methods have been applied in the past to reduce/eliminate the occurrence of SCC in these canisters. The effect of applied stresses during welding was studied by Wang et al. [11]. The authors reported that no cracking was observed when helium-doped materials were fusion welded under compressive loads of over 55 MPa, in contrast to cracking of 68% of total weld length when welded without any externally applied stresses. However, due to the high temperature and phase transformations occurring during fusion welding, the material is always susceptible to the formation of sigma phases and sensitization.

In recent years, solid-state mitigation techniques have been studied to improve the corrosion resistance of austenitic stainless steels and repair/mitigate cracking caused by environmental/chemical corrosion. Effect of low-temperature surface carburization on 4 point bend 304 austenitic stainless steel samples was reported to improve pitting corrosion resistance and SCC by inducing compressive stresses [12]. The use of compressive stresses to improve SCC resistance of austenitic stainless steels have also been shown using other solid-state technologies such as laser peening [13] and cold spray deposition. Yeom et al. [14] demonstrated the viability of cold spray deposition to mitigate and provide a physical barrier against chloride-induced stress corrosion cracking (CISCC). Recently, Ross et al. [15] and Gunter et al. [16] studied crack repair feasibility using friction stir processing (FSP). Ross et al. successfully repaired laboratory-generated CISCC cracks. Gunter et al. used FSP to repair EDM-tapered cracks growing up to widths of 2 mm. Both groups reported successful healing of these cracks after the experiments.

The goal of this study is to conduct friction surfacing experiments at spindle speeds up to 20,000 rpm with the motivation to reduce process forces while still sealing the crack. To the authors' knowledge, this regime of spindle speeds (>5,000 rpm) has not been reported in the literature. Friction surfacing experiments were conducted at spindle speeds of 2,000–4,000 rpm and the normal pressure during the process has been compared with tests at 20,000 rpm. The ability of friction surfacing to produce a fine-grained coating that can repair/mitigate cracking in 304L austenitic stainless steel has also been presented in this study. Friction surfacing experiments on simulated thru-cracks of width 50 μm were conducted and helium leak testing was conducted to evaluate the efficacy of friction surfacing deposit to produce an acceptable repair.

Experimental Setup

Friction surfacing experiments were conducted on a 3-axis computer numerically controlled (CNC) mill (HAAS, TM-1, Oxnard, CA, USA) for spindle speeds up to 4,000 rpm ("low speed") and on a 5-axis CNC mill-turn center (Mori-Seiki, NT1000W, Japan) for spindle speeds up to 20,000 rpm ("high speed"). All substrates were 3.5-mm-thick (0.1200-in-thick) 304L stainless steel sheet (North American Stainless, USA). The consumable rod (i.e., consumable tool) was also 304L stainless steel (Walsin Lihwa Corp., Taiwan). For "low speed" experiments, a 12.75-mm-diameter consumable rod was used, whereas a 5-mm-diameter consumable rod was used for "high speed" experiments. The 5-mm-diameter rod was used on the 5-axis CNC mill-turn center because of the maximum force limit of the tool spindle. Table 1 shows the chemical composition of the substrate and rod. The consumable

Table 1 Composition of 304L substrate and consumable rod

Element	304L standard*	Substrate	Consumable rod
	Weight%	Weight%	Weight%
Carbon	0.030 max	0.0152	0.02
Manganese	2.00 max	1.7400	1.68
Phosphorous	0.045 max	0.0310	0.033
Sulfur	0.030 max	0.0010	0.0278
Silicon	0.75 max	0.2730	0.33
Chromium	18.0–20.0	18.0100	18.23
Nickel	8.0–12.0	8.0610	8.09
Nitrogen	0.10 max	0.0939	0.0821
Molybdenum	–	0.3730	0.20
Copper	–	0.4670	0.56
Iron	Balance	Balance	Balance

* ASTM A240/240M–19

Table 2 Variables for friction surfacing experiments

Parameter	Spindle speed (rpm)	Diameter of consumable rod (mm)	Tool path
Low speed tests	20,00–40,00	12.75	Linear
High speed tests	20,000	5	Trochoid

rod was given an initial plunge (below the point of contact) of 1 mm at a constant plunge rate of 20 mm/min. The length of depositions was 40 mm. The rod was traversed along the substrate with a constant lateral traverse feed rate (V_X) and a constant axial (plunge) feed rate (V_Z). The workpieces were mounted atop a three-axis piezoelectric force dynamometer (Kistler model 9285), which measured the transient process forces. Regression analyses conducted on tests with varied V_X, V_Z, and spindle speeds showed that the normal process force (z-direction) was a strong function of the velocity ratio V_Z/V_X. This relationship has also been reported in previous studies [17]. In this study, all the data has been reported for V_Z/V_X of 0.33. It is to be noted that individual velocities V_X and V_Z were different for high and low speed tests. Both linear and trochoid tool paths were tested for friction surfacing experiments at 20,000 rpm. A trochoid is a curve traced by a point on a radius of a circle rotating along a straight line [18]. The trochoid tool path gave more stable and uniform depositions. Results with trochoid tool paths have been reported for experiments at 20,000 rpm. Variables for friction surfacing experiments have been summarized in Table 2.

For testing crack repair, substrate samples were prepared by machining faying surfaces and then tack welding (GTAW) two pieces together to create a sample with a simulated thru-crack with a width of approximately 50 μm (Fig. 2a, b) and overall sample dimensions of 38.1 mm × 25.4 mm × 3.5 mm (1.5″ × 1″ × 0.12″). Figure 2c shows a coating that was deposited on a sample with friction surfacing at 4,000 rpm spindle speed with a V_Z/V_X of 0.33. The overall sample dimensions are sized to fit on the helium leak tester (Pfeiffer Adixen ASM 142, France). The helium leak tester has a measurement range of 10^{-12}–10^{-5} Pa-m^3/s (10^{-11}–10^{-4} atm-cc/s or 7.5×10^{-12}–7.5×10^{-5} torr-l/s). For the measurements conducted, the base (background) leak rate was 6.5×10^{-11} Pa-m^3/s (6.5×10^{-10} atm-cc/s or 4.87×10^{-11} torr-l/s). All samples were cross sectioned at 2/3rd distance from the start of the deposit. For microstructural observation, the samples were ground, polished, and etched using an equal part solution of hydrochloric, nitric, and acetic acids. The

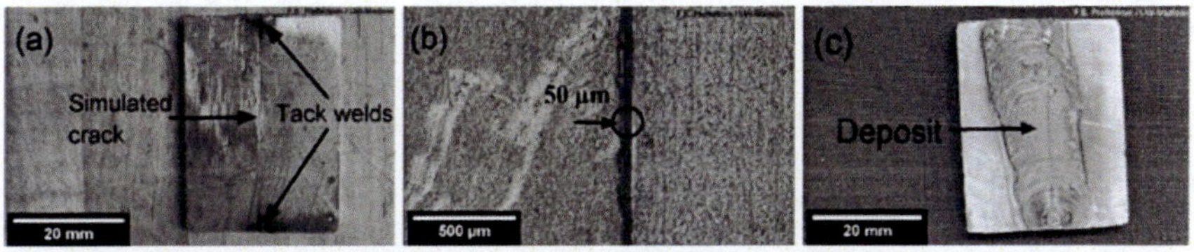

Fig. 2 Images showing **a** simulated crack sample, **b** width of simulated crack and **c** friction surfaced deposition on the sample

samples were observed and imaged using a white light optical metrology system (Alicona, InfiniteFocus® G4, Austria).

Results and Discussion

Friction surfacing deposits for low speed and high speed tests along with z-direction (normal) forces are shown in Fig. 3a, b, respectively. For the low speed tests with a linear tool path, the depositions are uniform and continuous with the width of deposition reducing and becoming constant as the steady-state regimen of the process is reached. For high speed depositions, the trochoid tool path leads to cyclical forces. The low speed deposition has an average steady-state force of approximately 5,000 N compared to 275 N for the high speed deposition.

Figure 4 shows optical images of cross sections of the substrate and consumable rod for friction surfacing at 4,000 rpm and 20,000 rpm spindle speeds. The deposits are not fully bonded to the substrate near the edges (Fig. 4a–c). For practical applications, there might be a requirement to machine off these un-bonded regions for better mechanical properties and corrosion resistance. Cross sections of the consumable rods reveal typical flash formation seen with friction surfacing (Fig. 4b, d). Grain size reduction is observed in deposited material when compared to the bulk grain structure of as received 304L austenitic stainless steel. The reduction in grain sizes for friction surfaced specimens at 4,000 rpm spindle speed have been shown in Fig. 5. This is a sign of dynamic recrystallization occurring due to intense plastic deformation and high strain rates during the process.

The average normalized width and thickness for four sets of spindle speeds used in this study have been summarized in Table 3. The normalization has been done with respect to the diameter of the consumable rod used for the tests. It is to be noted

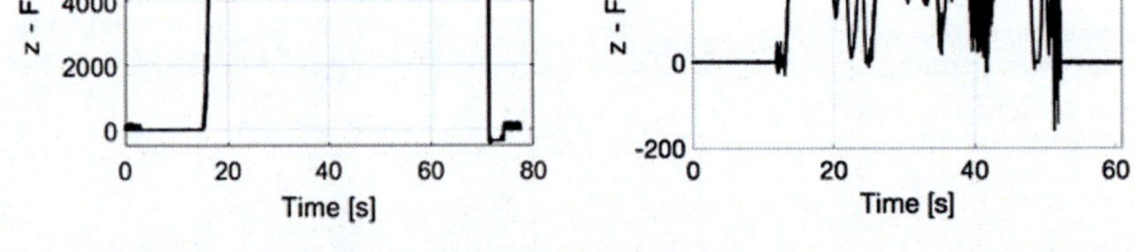

Fig. 3 Friction surfacing deposit and normal (z-direction) force plot for experiments at **a** 4,000 rpm and **b** 20,000 RPM

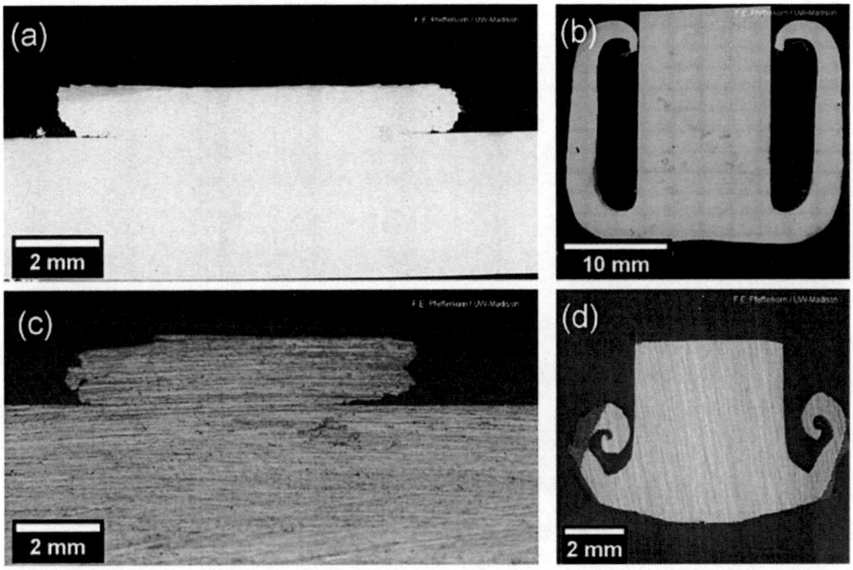

Fig. 4 Cross-sectional images of **a**, **c** substrate and **b**, **d** consumable rod after friction surfacing at 4,000 rpm (**a**, **b**) and 20,000 rpm (**c**, **d**)

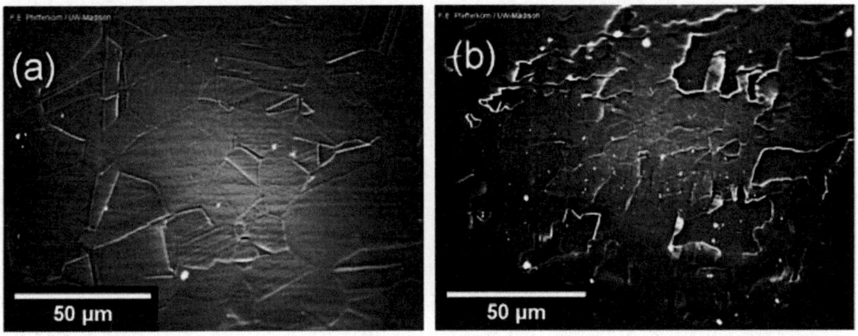

Fig. 5 Grain structure for austenitic stainless steel 304L **a** as received and **b** friction surfaced at 4,000 rpm spindle speed

Table 3 Average normalized width and thickness for friction surfacing experiments

Spindle speed (rpm)	Normalized width	Normalized thickness
2,000	1.16	0.19
3,000	1.09	0.13
4,000	1.01	0.07
20,000	0.82	0.25

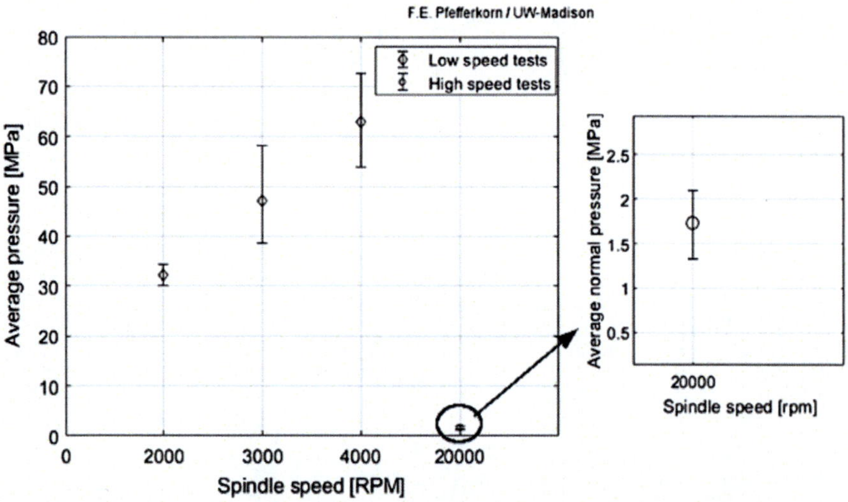

Fig. 6 Variation in average normal pressure across different spindle speeds

that the measured width includes un-bonded regions near the edge of the deposit. The width of deposition reduces with increase in spindle speeds; a phenomenon well reported in previous literature for friction surfacing and has been attributed to the reduction in the area of rotational contact plane between the consumable rod and the substrate [1, 19]. The surface of deposition is not smooth and contains ripples, characteristic with friction stir processes. Preliminary testing for bond quality was performed by attempting to mechanically peel off the coating using a chisel and hammer. For all the depositions reported in this study, the coating remained intact.

Figure 6 shows the variation in average normal pressure (z-direction) with spindle speed. The normal pressures have been compared instead of forces due to different diameter consumable rods used during low speed and high speed tests. As seen from the plot, average pressure increases with spindle speed up to 4,000 rpm; however, at 20,000 rpm the pressure drops significantly. This can be understood from the underlying physics of heat generation during friction surfacing. When the rotating consumable rod is in contact with the substrate, the heat generated is a combination of frictional heating and heat dissipation due to plastic deformation. The contribution of individual phenomena depends on the stick–slip factor, which itself is dependent on the tangential velocities of the rotating consumable rod. The fundamental equations for heat generation at the interface, variation of stick–slip factor as well as the coefficient of friction have been well established in the literature [20, 21]. As the spindle speed increases the contribution of frictional heating increases, and therefore a larger normal pressure is required to provide this frictional work. However, the self-limiting nature of friction surfacing limits the temperature rise of the material, with the maximum temperature achieved during the process being the solidus temperature of 304L stainless steel. Due to this self-limiting nature, once the peak frictional energy value is reached a further increase in spindle speed leads to a reduction in

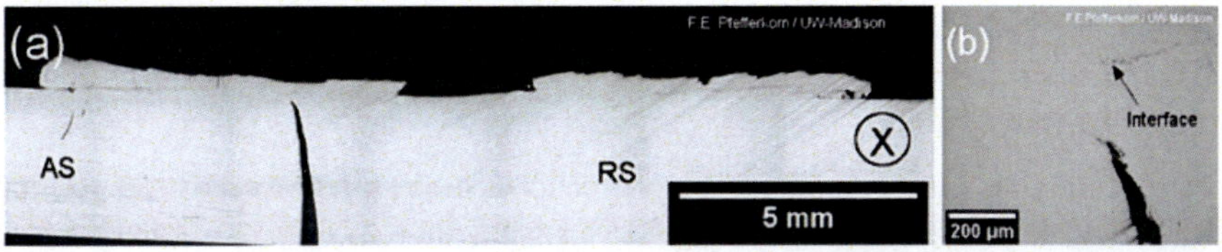

Fig. 7 Optical macrograph for friction surfaced deposit on simulated crack samples showing **a** complete cross section and **b** crack closing near the interface. The tool is travelling into the page

normal pressures, as seen for tests conducted at 20,000 rpm. It can be hypothesized that somewhere between 4,000 and 20,000 rpm a maximum pressure is achieved beyond which the normal pressures start reducing.

Figure 7 shows an optical image of a cross section of the sample with a simulated crack after friction surfacing. It can be seen that the thru-crack is closed near the interface of the coating and substrate. The simulated crack is closed to a depth of approximately 220 μm after friction surfacing (Fig. 7b). Bowing of the crack towards the advancing side (AS) of the deposition near the interface can be observed. It is to be noted that the tool is travelling into the page in Fig. 7. Friction surfacing introduces compressive stresses in the substrate close to the interface, along the cross section [8]. These compressive stresses can be contributing to the observed crack closure.

To evaluate the efficacy of friction surfacing deposits in sealing the crack, a helium leak test was conducted on sample shown in Fig. 2. After spraying helium gas on the deposit the base leak rate of the helium leak tester (6.5×10^{-11} Pa-m^3/s or 6.5×10^{-10} atm-cc/s or 4.87×10^{-11} torr-l/s) remained unchanged, i.e., no measurable helium leakage. This is an acceptable result based on ANSI N14.5 [22], which requires the helium leak rate to be less than 10^{-7} atm-cc/s (1.3×10^{-7} tor-l/s or 10^{-8} Pa-m^3/s). As shown in Fig. 5, friction surfacing also led to the formation of fine grain structure and dynamic recrystallization. Dynamic recrystallization has been reported to impart higher hardness and corrosion resistance in the coatings and at the interface. Preferential attack associated with the presence of delta-ferrites and sigma phases present in 304 stainless steel was studied by Rafi et al. [23]. They compared the reaction of as-received, friction surfaced, and gas tungsten arc welded 304 stainless steel samples to boiling nitric acid. Friction surfaced coatings performed superior to as received and GTAW samples due to the reduction/elimination of sigma phases.

Conclusions

Friction surfacing experiments with consumable rod and substrate made out of 304L stainless steel at 20,000 rpm spindle speed have been successfully demonstrated in this paper. The process pressures and deposition characteristics have been compared and contrasted with experiments conducted at low spindle speeds (2,000–4,000 rpm).

Application of friction surfacing as a mitigation and repair technique for austenitic stainless steel has also been demonstrated. The following conclusions can be obtained from the study:

- Conducting friction surfacing experiments at high spindle speeds (20,000 rpm) and with smaller diameter rods led to a significant reduction in normal pressure.
- The average pressure during the deposition increased from 2,000 to 4,000 rpm spindle speed and then reduced at 20,000. This variation can be attributed to the combined effect of the self-limiting nature of friction surfacing process and the influence of spindle speed on stick–slip characteristics during the process.
- Friction surfaced coatings repaired cracks in simulated samples to a depth of 220 μm below the interface of coating and substrate.
- Helium leak tests revealed no measurable leakage in samples with friction surfaced coating.

Considerable research is still to be completed with respect to generating a better understanding of thermomechanical events occurring during friction surfacing at higher spindle speeds. A more detailed characterization including measurement of residual stresses, grain orientations, and diffractography studies is required. Feasibility of friction surfacing as a crack repair/mitigation technique has been presented in this study. However, moving forward corrosion studies for these repairs will be conducted to ensure that the friction surfaced deposits don't sensitize in harsh environments. Mechanical testing of deposits will also be conducted to evaluate bond strength and hardness of coatings.

Acknowledgements The authors acknowledge the support of this work by the Department of Energy (grant DE-NE0008801), the Department of Mechanical Engineering, and the Department of Material Science and Engineering at the University of Wisconsin-Madison, the Machine Tool Technology Research Foundation, and colleagues in the Advanced Manufacturing Lab.

References

1. Gandra J, Krohn H, Miranda RM, Vilaça P, Quintino L, Dos Santos JF (2014) Friction surfacing - a review. J Mater Process Technol. https://doi.org/10.1016/j.jmatprotec.2013.12.008
2. Klopstock H (1941) An improved method of joining or welding metals. Patent specification Ref. 572789
3. Bedford GM, Vitanov VI, Voutchkov II (2001) On the thermo-mechanical events during friction surfacing of high speed steels. Surf Coat Technol 141
4. Rafi HK, Phanikumar G, Rao KP (2011) Material flow visualization during friction surfacing. Metall Mater Trans A 42(4):937–939
5. Lambrineas P, Jewsbury P (1992) Areal coverage using friction surfacing
6. Rafi HK, Babu NK, Phanikumar G, Rao KP (2013) Microstructural evolution during friction surfacing of austenitic stainless steel AISI 304 on low carbon steel. Metall Mater Trans A 44(1):345–350
7. Puli R, Ram GJ (2012) Dynamic recrystallization in friction surfaced austenitic stainless steel coatings. Mater Charact 74:49–54

8. Guo D, Kwok CT, Chan SLI (2019) Spindle speed in friction surfacing of 316L stainless steel–How it affects the microstructure, hardness and pitting corrosion resistance. Surf Coat Technol 361:324–341
9. Kanne WR Jr, Louthan MR Jr, Rankin DT, Tosten MH (1999) Weld repair of irradiated materials. Mater Charact 43(2–3):203–214
10. Park SHC, Sato YS, Kokawa H, Okamoto K, Hirano S, Inagaki M (2003) Rapid formation of the sigma phase in 304 stainless steel during friction stir welding. Scr Mater 49(12):1175–1180
11. Wang CA, Grossbeck ML, Aglan H, Chin BA (1996) The effect of an applied stress on the welding of irradiated steels. J Nucl Mater 239:85–89
12. Peng Y, Chen C, Li X, Gong J, Jiang Y, Liu Z (2017) Effect of low-temperature surface carburization on stress corrosion cracking of AISI 304 austenitic stainless steel. Surf Coat Technol 328:420–427
13. Lu JZ, Luo KY, Yang DK, Cheng XN, Hu JL, Dai FZ, Qi H, Zhang L, ZhongJS, WangQW, Zhang Y (2012) Effects of laser peening on stress corrosion cracking (SCC) of ANSI 304 austenitic stainless steel. Corros Sci 60:145-152
14. Yeom H, Dabney T, Pocquette N, Ross K, Pfefferkorn FE, Sridharan K (2020) Cold spray deposition of 304L stainless steel to mitigate chloride-induced stress corrosion cracking in canisters for used nuclear fuel storage. J Nucl Mater 152254
15. Ross K, Sutton B, Grant G, Cannell G, Frederick G, Couch R (2017) Development of friction stir processing for repair of nuclear dry cask storage system canisters. In: Friction stir welding and processing, vol IX. Springer, Cham, pp 39–46
16. Gunter C, Miles MP, Liu FC, Nelson TW (2018) Solid state crack repair by friction stir processing in 304L stainless steel. J Mater Sci Technol 34(1):140–147
17. Vitanov VI, Javaid N, Stephenson DJ (2010) Application of response surface methodology for the optimisation of micro friction surfacing process. Surf Coat Technol 204(21–22):3501–3508
18. Otkur M, Lazoglu I (2007) Trochoidal milling. Int J Mach Tools Manuf 47(9):1324–1332
19. Fukakusa K (1996) On the characteristics of the rotational contact plane - a fundamental study of friction surfacing. Weld Int 10(7):524–529. https://doi.org/10.1080/09507119609549043
20. Deng Z, Lovell MR, Tagavi KA (2001) Influence of material properties and forming velocity on the interfacial slip characteristics of cross wedge rolling. J Manuf Sci Eng 123(4):647–653
21. Nandan RGGR, Roy GG, Lienert TJ, Debroy T (2007) Three-dimensional heat and material flow during friction stir welding of mild steel. Acta Mater 55(3):883–895
22. ANSI N14.5-1997 (1998) Radioactive materials – leakage tests on packages for shipment. American National Standards Institute
23. Rafi HK, Phanikumar G, Rao KP (2013) Corrosion resistance of friction surfaced AISI 304 stainless steel coatings. J Mater Eng Perform 22(2):366–370

grain size from 40 to 5 μm due to dynamic recrystallization in the deposited layer. Reduction in grain size was also observed by Puli and Ram [7], who reported that the average grain size increased from 4.8 μm at the coating-substrate interface to 9.4 μm near the top of the coating. They attributed this change to the reduction in the cooling rate at the top of the coating. The authors also found necklacing of sub-grains around existing grains, which are a classic sign of discontinuous dynamic recrystallization. Discontinuous dynamic recrystallization was also observed by Guo et al. [8]; however, they reported the grains to be in competing domination between discontinuous and continuous dynamic recrystallization as spindle speed is changed. Changes in strain rate and temperatures during the process with changing spindle speeds were given as the reason.

Austenitic stainless steels have been widely used across various industries such as marine, nuclear, and transport for its high corrosion resistance. The grade 304L which is a low carbon grade (<0.03%) has been extensively used in the nuclear industry to make the pressure vessels for storing used nuclear fuel. Irradiation leads to n/α reactions with boron and nickel present in the stainless steels and forms helium [9]. Even though the low C content in austenitic stainless steels helps in the reduction of sensitization, the chromium-rich alloy leads to the formation of sigma and chi phases at prolonged exposure to elevated temperatures. These sigma phases reduce the corrosion resistance of 304L stainless steel [10]. Helium embrittlement attacks these grain boundaries rich in sigma phases and makes them susceptible to viscous effects such as pitting and stress corrosion cracking (SCC). Various methods have been applied in the past to reduce/eliminate the occurrence of SCC in these canisters. The effect of applied stresses during welding was studied by Wang et al. [11]. The authors reported that no cracking was observed when helium-doped materials were fusion welded under compressive loads of over 55 MPa, in contrast to cracking of 68% of total weld length when welded without any externally applied stresses. However, due to the high temperature and phase transformations occurring during fusion welding, the material is always susceptible to the formation of sigma phases and sensitization.

In recent years, solid-state mitigation techniques have been studied to improve the corrosion resistance of austenitic stainless steels and repair/mitigate cracking caused by environmental/chemical corrosion. Effect of low-temperature surface carburization on 4 point bend 304 austenitic stainless steel samples was reported to improve pitting corrosion resistance and SCC by inducing compressive stresses [12]. The use of compressive stresses to improve SCC resistance of austenitic stainless steels have also been shown using other solid-state technologies such as laser peening [13] and cold spray deposition. Yeom et al. [14] demonstrated the viability of cold spray deposition to mitigate and provide a physical barrier against chloride-induced stress corrosion cracking (CISCC). Recently, Ross et al. [15] and Gunter et al. [16] studied crack repair feasibility using friction stir processing (FSP). Ross et al. successfully repaired laboratory-generated CISCC cracks. Gunter et al. used FSP to repair EDM-tapered cracks growing up to widths of 2 mm. Both groups reported successful healing of these cracks after the experiments.

Real-Time Measurement of Friction Stir Tool Motion During Defect Interaction in Aluminum 6061-T6

Daniel J. Franke, Michael R. Zinn, Shiva Rudraraju, and Frank E. Pfefferkorn

Abstract The objective of this research is to develop a fundamental understanding of the interaction between features on the friction stir tool probe and volumetric subsurface defects formed during welding. This will guide the development of real-time defect monitoring methods that will promote process adoption in high volume and high-reliability applications. A single-head laser Doppler vibrometer system was used to produce a non-contact measurement of the eccentric motion of a friction stir tool during welding. When features on the tool probe interacted with voided volumes, the tool was momentarily deflected into the voided volume. The distortion signals in the tool position, measured with the laser vibrometer, are correlated with distortions in measured process forces and defect size. The results add understanding to the changes in forces signals that hold potential for defect monitoring and suggest that monitoring may be possible through a motion-based measurement (accelerometer) from the tool side.

Keywords Defects · Non-destructive evaluation · Tool motion · Laser doppler vibrometer

Introduction

Friction stir welding (Fig. 1) is a solid-state joining process that relies on mechanical intermixing of workpieces via plastic deformation at elevated temperatures. The mechanical intermixing is accomplished with a non-consumable rotating tool that consists of a larger diameter shoulder that resides at the workpiece surface and smaller diameter probe within the material thickness. The solid-state nature of the process provides advantages over fusion based welding processes. The advantages include a less severe heat affected zone, minimal distortion and residual stresses, avoidance of hot cracking, reduction/elimination of shielding gas, energy efficiency, and grain refinement within the stir zone due to dynamic recrystallization. A significant amount

---

D. J. Franke · M. R. Zinn · S. Rudraraju · F. E. Pfefferkorn (✉)
Department of Mechanical Engineering, University of Wisconsin-Madison, Madison, WI, USA
e-mail: frank.pfefferkorn@wisc.edu

Y. Hovanski et al. (eds.), *Friction Stir Welding and Processing XI*, The Minerals, Metals & Materials Series, https://doi.org/10.1007/978-3-030-65265-4_7

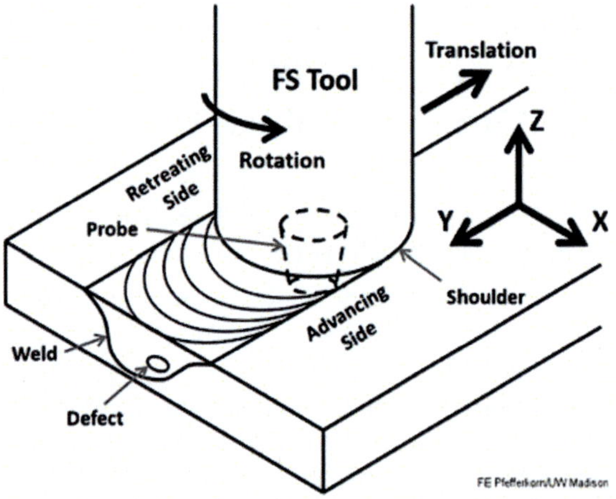

Fig. 1 Schematic of friction stir welding process showing the coordinate system used in the study

of research has shown that friction stir welding can be used as an energy-efficient method of creating high-quality joints in lightweight alloys such as aluminum and magnesium [1, 2].

However, process adoption has been limited by several aspects, one of which is the tendency for sub-surface defect formation due to incomplete material flow around the probe of the friction stir tool [3]. Sub-surface defects (that form down within the probe driven region) cannot be detected visually. In high-reliability applications, this requires a post-process non-destructive evaluation technique to determine if the weld is compromised, which in many cases makes the friction stir welding process cost-prohibitive. Furthermore, the friction stir welding process has been limited in high-volume production due to the constraint on process travel speed imparted by the potential for sub-surface defects to form at higher travel speeds. The objective of this work is to further the understanding of how sub-surface defects form during friction stir welding of aluminum alloys, as well as further the development of methods of real-time defect monitoring based on measured process outputs.

Prior research has shown that the flow of material around the tool probe occurs in an intermittent manner once per tool revolution [4–20]. This occurrence can be observed as layers of material within the microstructure formed at the distance of the advance of the tool per revolution. Additionally, significant research has shown that measured process forces tend to oscillate at the tool rotational frequency [11, 14–22]. This has led researchers to conclude that the intermittent flow of material once per revolution and the oscillation of force signals once per revolution are fundamentally linked [11, 14–22]. Researchers have proposed different explanations for the cause of the intermittent phenomena which include strain localization [7, 8], cavities forming and filling [13–17], thread driven flow [9], and eccentric motion of the tool (tool runout) [10–12, 20]. Researchers have focused on the intermittent nature of material flow around the tool probe because it has been shown that a breakdown in the intermittent flow of material tends to be the cause of sub-surface defects [12–20].

The current work seeks to further the understanding of the mechanisms surrounding the intermittent nature of material flow in the friction stir welding process. The specific objective is to test the hypotheses presented in Franke et al. [20] by examining the motion of the friction stir tool during welding. In [20], the researchers showed that the direction of the resultant component of the oscillatory process forces (within the X–Y plane of welding) points towards and rotates with the most eccentric point of the tool. This result suggests that the eccentric motion of the tool applies the oscillatory force to the workpiece. Additionally, it was hypothesized that momentary reductions in the oscillatory process forces during defect interactions are driven by momentary deflections of the friction stir tool from its nominally eccentric path. The goal is to further explain prior force measurements by adding a means of measuring the in situ motion of the friction stir tool. One device capable of measuring such a motion is a single point laser Doppler vibrometer (LDV). The laser Doppler vibrometer operates by measuring the velocity of a surface with a focused laser beam by using the Doppler shift between the incident light and scattered light returning to the measurement instrument. To the best of the authors knowledge, the only prior research that has examined the motion of the friction stir tool during welding was performed by Yan et al. [22] by means of a linear variable differential transformer (LVDT). The current work seeks to provide a deeper analysis of tool motion measurements with a focus on tool motion during defect interaction.

Experimental Methods

Apparatus

Friction stir welding was performed on a 3-axis CNC mill (HAAS TM-1). Two separate data acquisition systems were used to capture both the process forces and the motion of the tool during welding (Fig. 2). The data acquisition system used to measure the process forces that the tool applied to the workpiece is described in further detail in [20]. The system utilized a three-axis piezoelectric force dynamometer (Kistler model 9265) to measure the process forces in the three-axis coordinate system illustrated in Fig. 1. The process forces were measured in conjunction with the angular position of the friction stir tool via an angular encoder mounted on top of the milling machine spindle motor (HAAS, Part #: 30–30 390, 1024 pulses per revolution). The system provided the ability to measure net forces that the friction stir tool applies to the workpiece in conjunction with the angular position of the friction stir tool, thus allowing the angular position of features on the tool (e.g., flats, eccentric points) to be resolved at a given point in time with given force values. The in situ motion of the friction stir tool was captured during welding with a single point laser vibrometer system (Polytec, PSV-400 Scanning Head and Junction Box; PSV-A-420 Geometry Scan Unit; OFV-5000 Vibrometer Controller Unit). A friction stir (FS) tool made of heat treated H13 tool steel was utilized in this

Fig. 2 Image of the full experimental setup consisting of the force dynamometer fixture and laser vibrometer

study (Fig. 3). It consisted of a concave shoulder with a diameter of 15 mm and a threaded, conical probe with three flats. The probe diameter tapered from 7 to 5 mm and was 5 mm in length. The three flats on the tool probe were spaced 120° apart with all flats at a constant depth of 0.625 mm from the outer diameter of the probe. The probe was also threaded with a 1 mm pitch and a constant thread depth of 0.625 mm.

Vibrometer Setup

Welding on the HAAS TM-1 was performed by translating the workpiece and the fixture (attached to mill table) in the X–Y plane through a stationary rotating tool. This allowed the vibrometer laser to be aligned with the stationary tool. The vibrometer was positioned perpendicular to the weld travel direction on the advancing side of the weld as depicted in Figs. 2 and 4. One of the toe clamps was removed from the fixture on the advancing side in order to open up a line of sight from the vibrometer to the tool while the tool was engaged with the workpiece. The weld was programmed such that the table moves 130 mm in the Y direction to create a 130 mm weld. The line of sight was available between the two toe clamps labeled in Fig. 2, which allowed the vibrometer to capture the tool motion during the 90–105 mm section of the weld length. Prior to testing, the laser beam had to be aligned with the tool shoulder surface in three-dimensional space. Starting with a level vibrometer, the vertical position of the vibrometer was raised to align the beam just above the edge of the tool shoulder in the Z-direction. In the X-direction, the beam was autofocused on the tool surface and then manually focused 10 mm behind the tool surface to create a larger beam diameter at the tool surface. This allowed for an averaging effect over the larger beam diameter that more accurately captured the bulk motion of the tool as opposed to the micro-imperfections of the tool surface that a smaller beam diameter at peak focus would capture. The Y-position of the beam was adjusted while capturing data with a free-spinning tool until the velocity measurements oscillate around a value of zero.

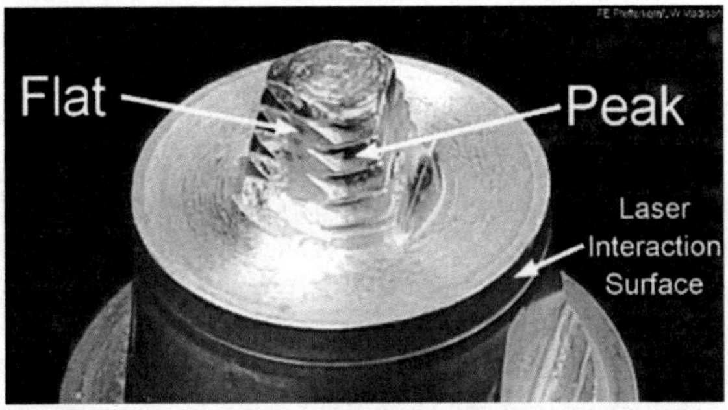

Fig. 3 Image of the friction stir tool used in the study showing the three flats on the tool probe opposed by the three peaks that interact with defects, as well as the surface of the tool the laser was focused on during welding

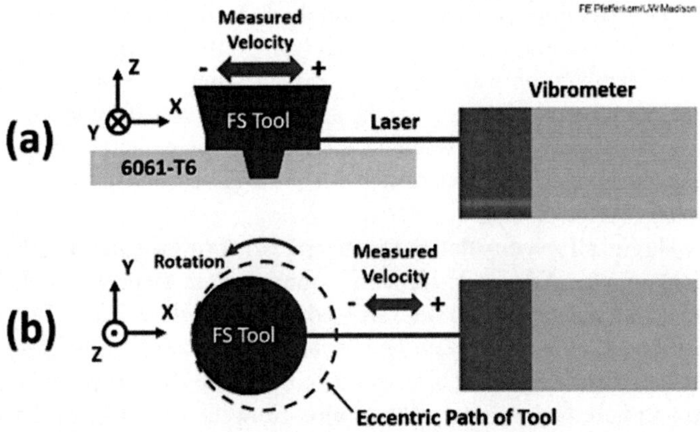

Fig. 4 Schematic of the laser vibrometer measurement **a** front view **b** top view

The measurement would produce velocity oscillations around an average positive or negative value if the beam was not centered on the tool in the Y-dimension. The Polytec PSV 400 velocity scale setting was set to "VD-07" (5 mm/s/V), all filters were turned off, and the bandwidth value was set to 0.5 kHz with 6400 FFT lines. These specific parameters allowed for 12.8 s of data to be captured at a sampling rate of 1.28 kHz. The acquisition settings are limited by the data buffer size and were selected to allow for a longer acquisition time in order to capture the full measurement region of interest along the 90–105 mm section of weld length.

Procedure

Both before and after welding, five total acquisitions of the free-spinning tool motion were collected while the tool rotated at 1,000 rpm (same as welding rotational rate). Four non-defective welds were performed at a rotational rate of 1,000 rpm and 200 mm/min, and eight defective welds (that resulted in sub-surface voids) were performed at 1,000 rpm and travel speeds ranging from 500 to 850 mm/min in increments of 50 mm/min. All welds were performed to a weld length of 130 mm, as a bead on plate weld, and with a 3-degree travel angle. An axial 0.2 mm plunge depth was set at the center of the shoulder for the four non-defective welds, and a plunge depth of 0.35 mm was set for the eight defective welds. The set negative values of plunge depth at the center of the tool shoulder result in the center of the tool shoulder residing near the top surface of the workpiece during welding due to the compliance of the system. All workpieces were 6061-T6 aluminum, and were 200 mm long, 100 mm wide and 6.35 mm thick. Prior to starting the force/encoder data acquisition system, the flat on the tool probe that the most eccentric point of the tool was centered on was positioned at the trailing edge of the tool. This set a zero point in the encoder data so that the angular position of the most eccentric point of the tool can be resolved throughout the welding force data. During welding, the vibrometer data acquisition was started when the laser reached the line of sight gap between the two toe clamps.

Post welding, all velocity data was converted to position data using the cumulative trapezoidal numerical integrator function “cumtrapz” in MathWorks MATLAB. The force and position data streams were aligned in the time domain by using the 105 mm weld position as a reference point. At 105 mm into the weld, the laser hits the fixture resulting in a momentary spike in the position data. The 105 mm point in time in force data is found based on a linear interpolation from the start time of the weld (spike in the travel direction force) to the end time of the weld (drop in the travel direction force). Five cycles of the corresponding force and position signals (taken around the 100 mm point in both data streams) were processed with a Discrete Fourier Transform to extract the amplitude of the third harmonic of the tool rotational frequency in both signals. Additionally, three cross-sections were cut from each weld at 100 mm into the weld length. The cross sections were polished, and the areas of the defects were measured using the optical technique used in [20].

Results and Discussion

Eccentric Motion of the Friction Stir Tool

The single point laser vibrometer measures the motion of the surface it is focused on in a singular dimension corresponding to the direction in which the beam is aligned.

The eccentric motion of the tool due to the natural tool runout occurs as a two-dimensional circular motion in the X–Y plane. Therefore, when the beam is aligned in the X-direction the circular two-dimensional eccentric motion of the tool is captured as a sinusoidal signal in the X-direction by the vibrometer. The amplitude of the free-spinning tool position signal captured by the vibrometer was examined in order to confirm that the vibrometer was capable of capturing the eccentric motion of the tool at the scale on which it occurs. The average peak-to-peak value of the position data was taken to be double the amplitude value at the tool rotational frequency extracted from the position signals by a Discrete Fourier Transform. The average of the peak-to-peak value from the tool position signal across all free spinning vibrometer tests was 76.1 ± 2.6 μm. An example of one revolution of the position signal measured from the tool spinning freely at 1,000 rpm is shown as the dotted line in Fig. 5. The small deviations in the signal from a perfect sinusoid are due to small inconsistencies of the shoulder surface from a perfect circle. The true runout of the tool (measured using a dial indicator) was 73.7 μm (0.0029 in), where the resolution of the dial indicator was 2.5 μm (0.0001 in). The total position change in the vibrometer signal is equal to the dial indicator measured true runout within the bounds of uncertainty for each measurement. Since the vibrometer confirmed the dial indicator measurement, it was assumed that that the vibrometer system can adequately capture the eccentric motion of interest.

The motion of the tool during a fully consolidated welding condition (when force signals oscillate at the tool rotational frequency without significant higher harmonics present, i.e., no defect interaction) is compared to the motion of the tool during the free spinning condition in Fig. 5. Across all fully consolidated welds performed, the amplitude of the position signal was on the order of 24% smaller than the free spinning position signal of the tool. This suggests the workpiece material constrains the eccentric motion of the tool during welding. This supports the results in Franke et al. [20], that showed that the direction of the oscillatory component of the process force that the tool applies to the workpiece in the X–Y plane tracks the most eccentric

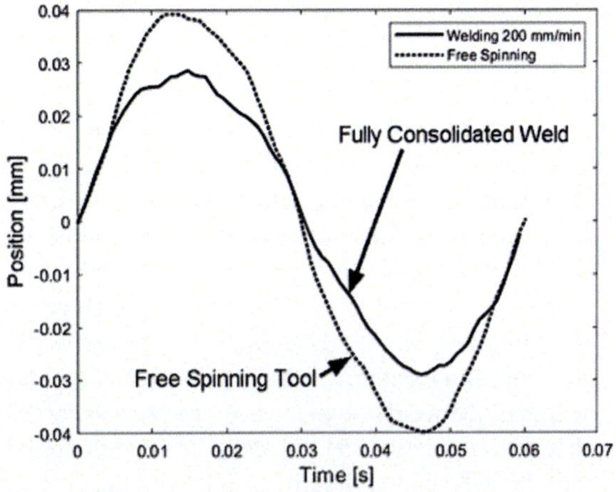

Fig. 5 A single revolution comparison of the position signal of a free spinning tool and a non-defective welding condition (1,000 rpm and 200 mm/min)

peak of the tool within each revolution (eccentric motion applies the force to the workpiece), and thus the equal and opposite reaction force should constrain the eccentric motion of the tool. This bolsters the conclusion that the eccentric motion of the tool due to tool runout is the primary driver of the oscillatory process forces. Prior results on tool motion measurements by Yan et al. [22] also showed small reductions in the periodic variations in tool position due to tool runout. The finding that the tool is constrained by the oscillatory processes forces refutes the hypothesis by Fonda et al. [10] that proposed that oscillatory process forces lead to the propagation of eccentric tool motion, which amplifies the effect of the eccentric motion of the tool on the intermittent flow of material around the tool probe. However, the results still support the primary part of the hypothesis proposed by Fonda et al. that the eccentric motion of the tool is the primary driver of the intermittent flow of material. The only discrepancy is the propagation of motion.

Tool Motion During Defect Interaction

In prior work published in Franke et al. [20], the primary hypothesis focused on the physical explanation of what occurs within one tool revolution when features on the tool probe (peaks between flats) interact with subsurface volumetric defects and alter the process force transients. It was hypothesized that the force that the eccentric motion of the tool applied to the workpiece was momentarily reduced when the peaks of the tool probe interacted with defective volumes because the tool was momentarily deflected from its eccentric path into the defective volumes. This hypothesis was tested by measuring the motion of the friction stir tool with the laser vibrometer during defect interaction. During a fully consolidated welding condition, the process forces oscillate primarily at the tool rotational frequency and the corresponding position signal is also primarily a sinusoid at the tool rotational frequency as shown in Fig. 6a, b. However, when welding in a defective regime, two of the peaks on the tool probe interact with the defective volume formed each revolution to produce two distortions in the force and position signals as shown in Fig. 6c, d. For both the force and position signals, the positive X-direction is towards the advancing side of the weld (Fig. 1). When the most eccentric point of the tool was towards the advancing side of the weld, one of the peaks on the probe opposite the most eccentric point interacted with the defective volume that had started to form on the retreating side of the weld. The narrow peak on the tool probe allowed the tool to be momentarily deflected towards the retreating side into the defective volume. This is labeled as Interaction 1 in Fig. 6c, d. The deflection of the tool towards the retreating side momentarily reduced the force that the eccentric motion of the tool applied to the workpiece on the advancing side. This deflection was resultant of the relaxing of the tool from its nominal position in the X direction that the tool resided at along the length of the weld due to the average process force in the negative X-direction. Note that in the defective welding data shown, the average process force in the X-direction was 2,000 N, which should result in significant steady-state deflection of the tool due

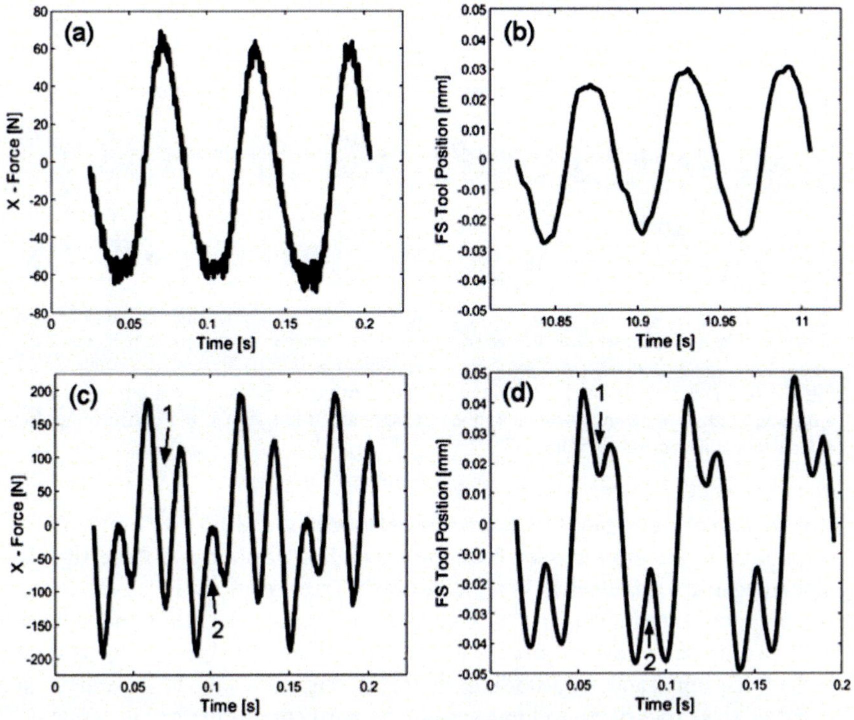

Fig. 6 Three revolutions of **a** X-direction force fully consolidated weld, **b** X-direction tool position fully consolidated weld, **c** X-direction force defective weld, **d** X-direction position defective weld. Note: the average component of the force signals have been removed, i.e., normalized around zero

to the compliance of the system in the X-direction (estimated to be 2 MN/m [20]). The second interaction (Interaction 2 in Fig. 6) occurred when a second peak on the tool probe moved into the defective volume on the advancing side of the weld which caused the tool to momentarily deflect towards the advancing side causing a momentary reduction in the force that the eccentric motion of the tool applied to the workpiece on the retreating side. The angular encoder data was utilized to verify that that the two distortions within the two data streams align with each other in the time domain. The angular position of the most eccentric point of the tool is known in the encoder/force timing scheme. Therefore, the point in time in the encoder/force data when the most eccentric point of the tool is at the advancing side edge (nearest to the vibrometer) can be approximately aligned to the most positive point in time in the vibrometer position signal.

Prior research on the force transients (generated by the interactions previously described) has shown that the distortions in the force signals can be extracted as an amplitude at the third harmonic of the tool rotational frequency (the two distortions generate three peaks per revolution) [17–20]. For the corresponding force and positions signals for each given weld, the amplitude at the third harmonic of the

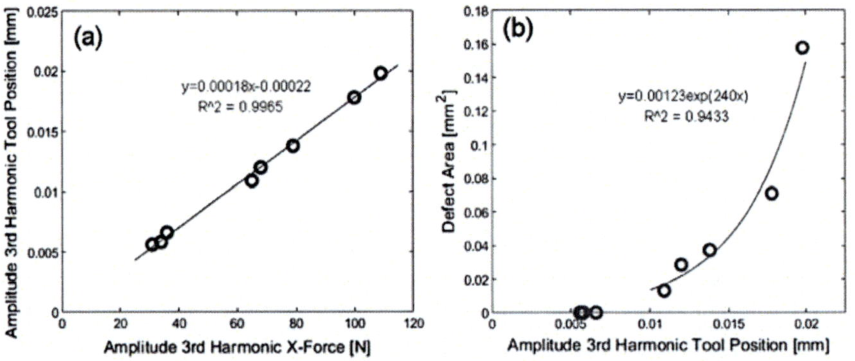

Fig. 7 Relationship between the amplitude of the third harmonic in the tool position signal and **a** the amplitude of the third harmonic in the corresponding force signal, and **b** the average area of defects taken from cross sections

tool rotational frequency was extracted using a Discrete Fourier Transform. The amplitudes at the third harmonic for corresponding position and force signals are plotted against each other in Fig. 7a. A direct linear relationship can be observed, which supports the fundamental linking of the momentary deflection and momentary reduction in the oscillatory process forces. The relationship between the distortions in the force signal and the distortions in position signals should be dependent on the dynamic properties of the welding system, primarily the stiffness of the system. The stiffness of the system used for friction stir welding can vary significantly from more compliant robotic arms to more rigid dedicated gantry-style systems. How altering the stiffness of the machine alters the force and deflection relationship must be investigated further. The stiffness in the X-direction of the system used in this study was estimated to be on the order of 2 MN/m [20].

Prior research has shown that a direct relationship can be drawn between the distortions in force signals and the size of sub-surface defects left within the weld [17–20]. The result that the momentary deflection of the tool is fundamentally linked to the distortions in the forces signals suggests that a relationship between the distortions in the tool position signal and defect size can be developed. Figure 7b shows the relationship between the amplitude of the third harmonic in the friction stir tool position signal and the average defect area taken from the cross-sections of each of the corresponding welds. The trend is very similar to the relationship between the amplitude at the third harmonic in the force signal and defect size reported in [20]. The amplitude of the third harmonic in both signals increased with defect size but reached a limit where the amplitude saturates above a certain defect size. The relationship between the distortion in the tool position signal and defect size suggests that a motion-based measurement (accelerometer) has potential to form the basis of a method of defect monitoring. An accelerometer-based measurement may be easier to implement from the tool side of the process in a production setting as compared to a force-based measurement. Future examinations must determine if an accelerometer can capture similar tool motions.

Conclusions

A laser vibrometer system was utilized to produce a real-time non-contact measurement of the microscale motions of a friction stir tool during welding of aluminum 6061-T6 in both fully consolidated and sub-surface defective welding regimes. The conclusions that can be drawn from the results are:

- In fully consolidated welds, during which the tool was in complete contact with the workpiece material (no void interactions that we know of), the workpiece material constrains the eccentric motion of the tool as compared to its eccentric motion due to its runout when it was spinning freely. This supports previous research that suggested that the eccentric motion of the tool was the primary driver of oscillatory process forces in the plane of welding.
- In defective welding regimes, when the peaks (created by flats) on the tool probe interacted with subsurface volumetric defects, the tool was momentarily deflected into the defective volume. This supports a previous hypothesis that described the physical nature of changes in process force transients during defect interaction.
- The momentary deflection of the tool during defect interaction suggests that a motion-based measurement (accelerometer) holds potential as the basis of a defect monitoring system.
- The stiffness of the welding system appears relevant to the development of a defect monitoring method based on either a motion-based or force-based measurement of the interactions described.

Future work will focus on developing a dynamic model in order to mathematically describe the motion of the tool for a given machine properties (mass, stiffness, and damping), as well as examining how different aluminum alloys alter the process forces and tool motions during sub-surface defect interaction. Additionally, the efficacy of using an accelerometer-based measurement as the basis of a defect monitoring system will be examined.

Acknowledgements The authors gratefully acknowledge financial support of this work by the Department of Mechanical Engineering at the University of Wisconsin-Madison and the U.S. National Science Foundation through grant CMMI-1826104. The authors would also like to acknowledge Prof. Melih Eriten and Lijie Liu for loaning and guidance on the use of the laser vibrometer system, which was enabled through the National Science Foundation MRI Grant CMMI 1725413.

References

1. Mishra RS, Ma ZY (2005) Friction stir welding and processing. Mater Sci Eng R 50:1–78
2. Threadgill PL et al (2009) Friction stir welding of aluminium alloys. Int. Mat. Rev. 54:49–93
3. Mishra D, Roy RB, Dutta S, Pal SK, Chakravarty D (2018) A review on sensor based monitoring and control of friction stir welding process and a roadmap to Industry 4.0. J Manuf Processes 36:373–397. https://doi.org/10.1016/j.jmapro.2018.10.016

4. Krishnan KN (2002) On the formation of onion rings in friction stir welds. Mater Sci Eng A 327:246–251
5. Colligan K (1999) Material flow behavior during friction stir welding of aluminum. Weld Res Suppl 229–237
6. Prangnell PB, Heason CP (2005) Grain structure formation during friction stir welding observed by the 'stop action technique.' Acta Mater 53:3179–3192
7. Schneider JA, Nunes AC (2004) Characterization of plastic flow and resulting microtextures in a FSW. Metall Mater Trans B 35(4):777–783
8. Nunes AC (2006) Metal Flow in Friction Stir Welding. Materials Science & Technology, (2006) Conference and Exhibition, October 15–19, 2006. Cincinnati, Ohio, USA
9. Chen ZW, Cui S (2008) On the forming mechanism of banded structures in aluminum alloy friction stir welds. Scripta Mater 58:417–420. https://doi.org/10.1016/j.scriptamat.2007.10.026
10. Fonda R, Reynolds A, Feng CR, Knipling K, Rowenhorst D (2013) Material flow in friction stir welds. Metall Mater Trans A 44:337–344
11. Reynolds AP (2008) Flow visualization and simulation in FSW. Scripta Mater 58:338–342. https://doi.org/10.1016/j.scriptamat.2007.10.048
12. Gratecap F, Girard M, Marya S, Racineux G (2012) Exploring material flow in Friction stir welding: tool eccentricity and formation of banded stuctures. Int J Mater Form 5:99–107. https://doi.org/10.1007/s12289-010-1008-5
13. Abergast WJ (2008) A flow-partitioned deformation zone model for defect formation during friction stir welding. Scripta Mater 58:372–376
14. Boldsaikhan E, Burford DA, Gimenez P (2011) Effect of plasticized material flow on the tool feedback forces during friction stir welding. Friction Stir Welding and Processing VI, 335–343
15. Boldsaikhan E, Corwin EM, Logar AM, Arbegast WJ (2011) The use of neural network and discrete fourier transform for real-time evaluation of friction stir welding. Appl Soft Comput 11:4839–4846
16. Boldsaikhan E, McCoy M (2013) Analysis of tool feedback forces and material flow during friction stir welding. Friction Stir Welding and Processing VII, 311–320
17. Shrivastava A, Zinn M, Duffie NA, Ferrier NJ, Smith CB, Pfefferkorn FE (2017) Force measurement-based discontinuity detection during friction stir welding. J Manufact Process 26:113–121
18. Shrivastava A, Pfefferkorn FE, Duffie NA, Ferrier NJ, Smith CB, Malukhin K et al (2015) Physics-based process model approach for detecting discontinuity during friction stir welding. Int J Advanced Manuf Technol 79:604–615. https://doi.org/10.1007/s00170-015-6868-x
19. Franke DJ, Zinn MR, Pfefferkorn FE (2019) Intermittent flow of material and force-based defect detection during friction stir welding of aluminum alloys. In: Hovanski Y, Mishra R, Sato Y, Upadhyay P, Yan D (eds) Friction stir welding and processing X. The minerals, metals & materials series, Springer, Cham, p 149–160. https://doi.org/10.1007/978-3-030-05752-7_14
20. Franke DJ, Rudraraju S, Zinn MR, Pfefferkorn FE (2020) Understanding process force transients with application towards defect detection during friction stir welding of aluminum alloys. J Manuf Proc 54:251–261. https://doi.org/10.1016/j.jmapro.2020.03.003
21. Li WY, Li JF, Zhang ZH, Gao DL, Chao YJ (2013) Metal flow during friction stir welding of 7075–T651 aluminum alloy. Exp Mech 53:1573–1582. https://doi.org/10.1007/s11340-013-9760-3
22. Yan JH, Sutton MA, Reynolds AP (2007) Processing and banding in AA2524 and AA2024 friction stir welding. Sci Technol Weld Joining 12(5):390–401. https://doi.org/10.1179/174329307X213639

Preliminary Investigation of the Effect of Temperature Control in Friction Stir Welding

Johnathon B. Hunt, David Pearl, Yuri Hovanski, and Carter Hamilton

Abstract Friction stir welding (FSW) is an advantageous solid-state joining process, suitable for many hard to weld materials in the energy, aerospace, naval, and automotive industries. Precipitation strengthened alloys, specifically 2XXX and 7XXX series aluminum alloys, are often joined by FSW to protect the strength of the materials and to avoid cracking. To maximize the strength of FSW joints in these precipitation hardened alloys, the thermal input affect must be better understood. The authors hypothesised that controlling the welding temperature under the dissolution temperature would result in stronger joints. To test the main hypothesis single alloy friction stir "butt" welds were produced, from aluminum 2024-T351 and 7075-T651 alloys, and tensile tested. Spindle speed proportional–integral–derivative (PID) temperature control was implemented to achieve sub-dissolution welding temperatures. This preliminary study will supply additional research to better understand the resulting microstructure, weld properties of sub-dissolution FSW. In addition, a numerical simulation to represent the temperature distribution will be built. Then optimized FSW temperatures could be predicted and tested in these alloys.

Keywords Friction stir welding · Temperature control · 2024 alloy · 7075 alloy

Introduction

Friction stir welding (FSW) is an advantageous solid-state material joining process used in the automotive, naval, aerospace, and energy industries that was invented by The Welding Institute [1, 2]. The process includes two work pieces that have a seam between them. A FSW tool, comprising of a cylindrical shoulder and a concentric pin, often threaded, is plunged down into the seam while rotating. Once the shoulder is engaged with the workpiece, the tool will advance along the seam while rotating,

J. B. Hunt (✉) · Y. Hovanski
Brigham Young University, Provo, UT, USA
e-mail: john.hunt92@gmail.com

D. Pearl · C. Hamilton
Miami University, Oxford, OH, USA

Y. Hovanski et al. (eds.), *Friction Stir Welding and Processing XI*, The Minerals, Metals & Materials Series, https://doi.org/10.1007/978-3-030-65265-4_8

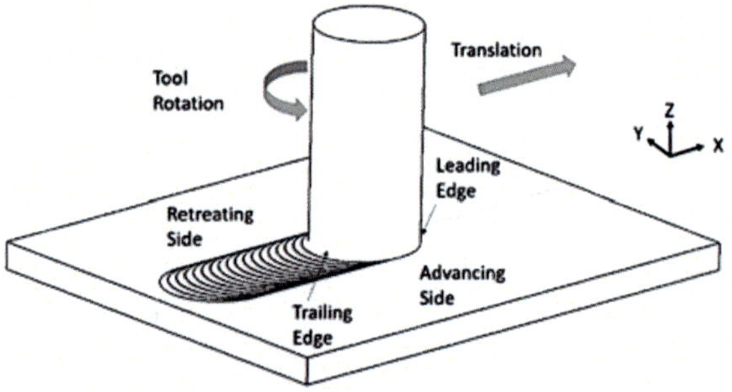

Fig. 1 A schematic of FSW and the coordinate system used in this work

see Fig. 1. The friction between the tool and work piece creates heat and plastic deformation which results in a consolidated joint between the two sheets. There are three main advantages of FSW: first, the energy required to produce a friction stir weld regularly is less than the energy required to produce a fusion weld [3]; second, friction stir welded joints can have superior material properties in comparison to traditional fusion welded joints [4]; finally, FSW can be less expensive because it is more environmentally friendly due to the lack of consumable products during the process, such as shielding gas or electrodes [5].

FSW also avoids hot cracking common in traditional welding methods of 2xxx and 7xxx series aluminum alloys [6, 7] because of the lower welding temperatures [8, 9]. The feasibility of welding these alloys is of great import to many industries, and research about residual stresses [10, 11], microstructure [12–14], fatigue [15], corrosion [16] and others properties of friction stir welded joints have been studied. Combined, this body of research provide the knowledge to produce high performance welds [17]. However, implementing this knowledge to manufacture superior welds is not trivial [18–20]. The correct use of tool design, process control, machine parameters, weld set up, backing plate and others are necessary to create high quality welds. This paper will focus on one of these FSW tools, specifically temperature control.

Temperature control for FSW has been developed since 2007 [21]. This body of research includes two main focus points: power or torque based temperature control [21–25] and spindle speed based temperature control [26–30]. Most of these temperature control methods are based on a predictive thermal model [31] or PID gains [32] to control temperature throughout a weld. Both methods are very capable to hold a weld within plus or minus 2 degrees Celsius [33].

This paper provides the results from a preliminary study focused on optimizing friction stir welded joints of 2024-T351 and 7075-T651 using temperature control. The authors hypothesised that if the weld temperature were controlled beneath the dissolution temperatures of these materials, more elevated-temperature strengthening phases (such as η'/η for 7075 and S'/S for 2024) would remain in the weld nugget after processing, thus providing a stronger joint.

Methodology

Two sets of single alloy friction stir "butt" welds, produced from precipitation strengthened aluminum 2024-T351 [34] and 7075-T651 [35], with dimensions of 609.6 mm × 107.95 mm × 6.35 mm, supplied the data for this study. All welds were performed at room temperature. Figure 2 depicts the set up for a weld. Position control was implemented for all of the welds with a spindle speed PID temperature controller. PID gains were obtained by using an adaptive feedback relay test [32] on an TTI model RM-2, B&R controller, Bond supported, FSW machine on Brigham Young University campus. The maximum spindle speed allowed for the 7075 alloy was 200 RPM and 300 RPM for the 2024 alloy. The three setpoint temperatures are found in Table 1. A replicate weld was produced for each setpoint temperature resulting in 6 total welds for each alloy. The temperatures between the two materials differ because parameter verification in the 2024-T351 aluminum alloy revealed defects in the weld when controlled at 390 °C. Thus, the control temperatures for the 2024 material were both raised to keep a 50 °C temperature difference between the two.

Other welding parameters for of all of the welds included a traverse speed of 150 mm/min and a 3° backward tilt. The method that the material was constrained is shown in Fig. 2. Dowel pins and set screws constrained the material in plane to the welding direction, and 3 toe clamps with a steel bar constrained the material downward. A 6.365 mm thick mild steel plate was used as backing between the workpieces and welding anvil. To avoid the "lazy S" FSW defect, the welders roughly ground off the oxide layer 25.4 mm wide on the seam of the material Fig. 2. The FSW tool was a H13 tool steel CS4-PB06 [36] with a convex, four start 1.25 rotation

Table 1 Displays all of the different temperatures that each material was welded. There were two welds per temperature

7075	2024
Uncontrolled	Uncontrolled
430 °C	440 °C
390 °C	410 °C

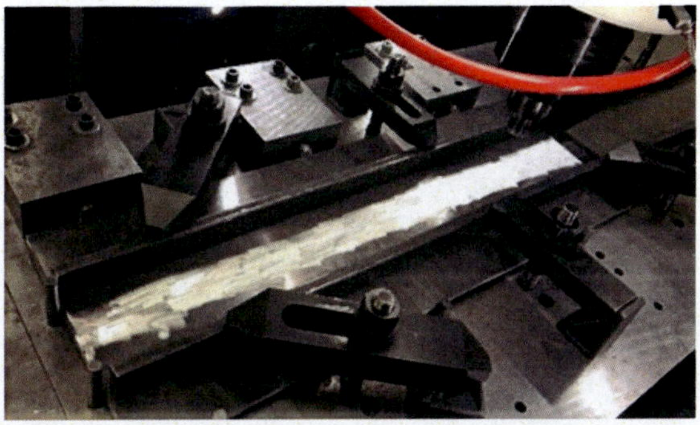

Fig. 2 Is an example of the weld setup for all temperature control welds. (Color figure online)

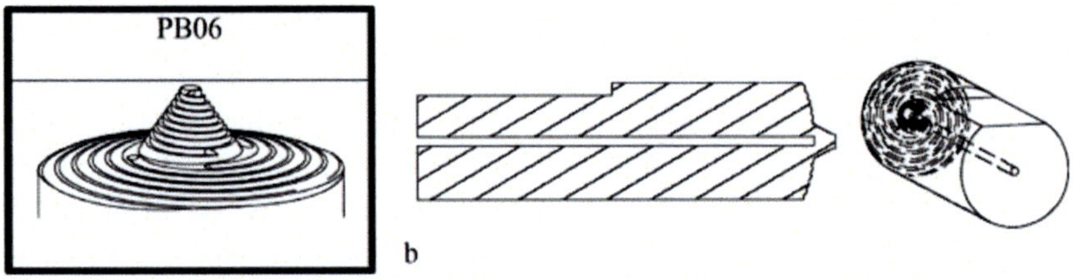

Fig. 3 Includes **a** visual of the FSW **b** tool geometry and **c** location of the thermocouple hole

scrolled shoulder, and 25.27 mm in diameter shaft. The threaded pin includes a 37° taper from a top diameter of 1.78 mm to a base diameter of 7.62 mm and a length of 5.08 mm. Figure 3 offers a visual of the FSW tool geometry. This tool welded both sets of alloys. A sodium hydroxide solution cleaned the tool in between welding the different materials. To measure a relative temperature, a thermocouple hole was EDM drilled in the center of the tool down to 0.25 mm from the center of the pin as shown in Fig. 3. A "K type" thermocouple connected to a Lord MicroStrain, TC-LINK-200-OEM chip and a Bluetooth receiver, WSDA-BASE-101_LXRS to interfaced with the analog data acquisition native to the TTI FSW machine. The x, y and z force as well as torque data were obtained from three Kistler tri-directional load cells that are placed 120 degrees apart in between the frame of the machine and the tool. All signals were measured at a rate of 1250 Hz.

The authors waited 1700 h before obtaining tensile tests data. ASTM E8 size dog bones were waterjet cut out of the welds. The authors acknowledge that the use of a waterjet may induce a "skirting" effect that might slightly increase the cross sectional area of the tensile specimens, however the waterjet cuts were necessary to ensure that the microstructure did not change in the cutting process. An MTS Exceed tensile testing machine, Model E44, loaded the dog bones axially until failure at a constant velocity of 0.2 mm/sec. Position was measured with a MTS extensometer, model 634.12E-54. Averaged ultimate strength values from the beginning, center and end of the weld were calculated by taking three dog bones ultimate strengths within 30 mm apart from each other.

Results and Discussion

The resulting temperatures for all twelve welds can be found in Fig. 4. The PID temperature control was executed after the tool had plunged into the material and finished the traverse travel velocity ramp. Thus, temperatures stabilized near 100 mm from the start of the weld. The uncontrolled temperatures did not significantly rise as expected. This could be caused from the faster translational speed and slower spindle speed in comparison to Khodir and Guo's work [8, 10] who performed welds at 50 mm/min and lowest spindle speed of 400 RPM. Faster velocity and slower spindle speed would allow for greater heat dissipation after the FSW tool has past a

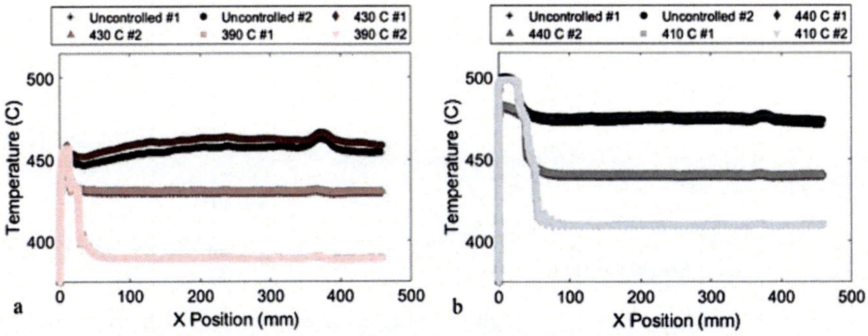

Fig. 4 Includes the resulting temperatures of all twelve welds **a** aluminum 7075-T651 alloy **b** aluminum 2024-T351 alloy. (Color figure online)

certain point resulting in a more consistent temperature throughout the weld. Even though the uncontrolled temperatures in this study were stable, it would be extremely difficult to produce a defect-free weld with a single low spindle speed from the start to finish. This can be demonstrated in Fig. 5 where the spindle speed for all of the welds are plotted.

Using both Figs. 4 and 5, the requirement to have a high enough spindle speed and plunge time to acquire a sufficient temperature can be understood. This type of plunge builds to a temperature that sufficient to dissolve the precipitates in both alloys, 450 °C for 7075 and 500 °C for 2024 in addition to the mechanical energy that the tool introduces. This dissolution softens that material and allows the FSW tool to enter that material. Once the FSW tool is in the material, it can proceed with the FSW process. The resultant thermo-mechanical input and cooling rates are then defined by the welding parameters thereafter. To acquire the desired temperatures the PID temperature controller, generally, reduced the spindle speed. Once the desired temperature was achieved smaller adjustments were required to hold instead of wandering to higher temperatures. In this study the coldest setpoint temperatures

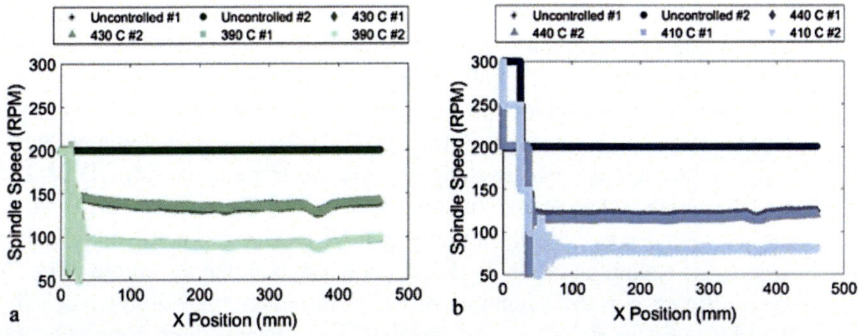

Fig. 5 Conveys the spindle speed required to hold the setpoint temperatures **a** aluminum 7075-T651 alloy **b** aluminum 2024-T351 alloy. (Color figure online)

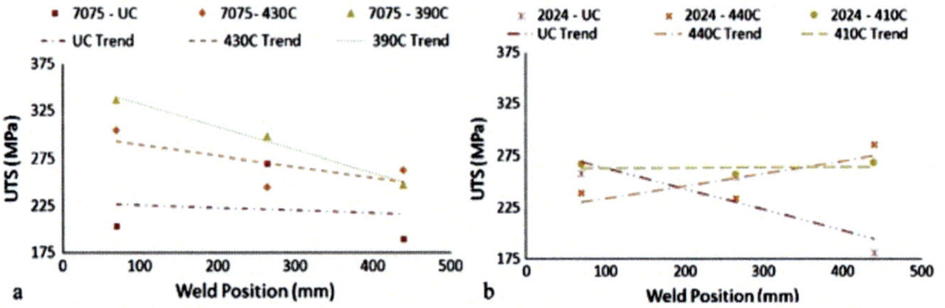

Fig. 6 Contains averaged ultimate strength of the beginning, center and end of the weld for **a** 7075-T651 aluminum alloy **b** 2024-T351 aluminum alloy. (Color figure online)

for both alloys had the highest ultimate strengths among the different temperature regimes Fig. 6.

The authors hypothesize, from the 7075 strength data in Fig. 6, that the higher strengths from the lower setpoint temperatures were caused from the aging that occurred after the dissolution of the plunge. This hypothesis would also support the decrease in strength later in the weld, where the strength decreases as a result of the lack of dissolution and aging. Instead, the rest of the weld over ages at the welding temperatures. The 2024 material does not follow this trend however. Only the uncontrolled regime has a continual decrease in strength, while the temperature controlled welds had a more consistent strength or increase in strength. The authors believe it possible that the setpoint temperatures and welding parameters were sufficient to dissolve the precipitates and provide a cooling rate that enabled growth of new precipitates (*GP* zones) upon cooling post-processing. The future body of research that is being conducted will be helpful in accepting or rejecting these hypotheses.

Conclusion

(1) This study demonstrates that temperature control FSW under the dissolution temperatures of aluminum alloys 2024 and 7075 has a clear effect on the resulting properties of the weld. Specially, a significant decrease in strength was observed in both alloys from the beginning to the end of the weld with certain temperature control regimes. The 7075-T651 alloy, welded at 390 °C, had a 39.7% strength decrease along the weld, while the uncontrolled 2024-T351 weld, near 475 °C had a strength decrease of 32.4%. The authors postulated that this strength decrease is a result of respective precipitates being dissolved and aged at the beginning of the weld, but not throughout the rest of the weld. The main difference at the beginning of the weld is the temperatures reached by the plunge of the FSW tool. As the FSW tool plunges, high enough temperatures to dissolve respective alloy's precipitates are achieved with a cooling effect once the tool moves on to weld the material seam. This combination of

events provides a precipitation heat treatment for the beginning of the weld, thus resulting in a higher strength. The rest of the weld did not achieve dissolution temperatures, and the weld overaged resulting in lower strengths.

(2) One setpoint temperature in the 2024-T351 alloy achieved stable strength throughout the entire weld, with only a standard deviation of 6.4 MPa. This data suggest that an appropriate thermo-mechanical input would result in desired weld properties consistent throughout the whole weld.

(3) The affect that sub-dissolution FSW temperature control has on weld properties is not trivial. The type of plunge, welding parameters, weld setup as well as setpoint temperature all contribute to the resultant weld properties. Future study need to be conducted to solidify the hypotheses made in this work.

References

1. Thomas WM, Nicholsa ED, Needham JC, Murch MG, Temple-Smith P, Dawes CJ (1995) Improvements Relating to Friction Stir Welding, United Kingdom [Online]. Available: https://data.epo.org/gpi/EP0615480A1-IMPROVEMENTS-RELATING-TO-FRICTION-WELDING.html
2. Thomas WM, Nicholas ED (1997) Friction stir welding for the transportation industries. Mater Des 18(4):269–273. https://doi.org/10.1016/S0261-3069(97)00062-9
3. Lohwasser D, Chen Z (2009) Friction Stir Welding: From Basics to Applications, Woodhead Publishing Limited
4. Mishra RS, Ma ZY (2005) Friction stir welding and processing. Mater Sci Eng R 50(1):1–78. https://doi.org/10.1016/j.mser.2005.07.001
5. BT Gibson et al (2014) Friction stir welding: process, automation, and control. J Manuf Processes 16(1):56–73. https://doi.org/10.1016/j.jmapro.2013.04.002
6. Holzer M, Hofmann K, Mann V, Hugger F, Roth S, Schmidt M (2016) Change of hot cracking susceptibility in welding of high strength aluminum alloy AA 7075. Physics Procedia 83:463–471. https://doi.org/10.1016/j.phpro.2016.08.048
7. Ghaini FM, Sheikhi M, Torkamany MJ, Sabbaghzadeh J (2009) The relation between liquation and solidification cracks in pulsed laser welding of 2024 aluminium alloy. Mater Sci Eng A 519(1):167–171. https://doi.org/10.1016/j.msea.2009.04.056
8. Khodir SA, Shibayanagi T, Naka M (2006) Microstructure and mechanical properties of friction stir welded AA2024-T3 aluminum alloy. Mater Trans 47(1):185–193. https://doi.org/10.2320/matertrans.47.185.
9. Cavaliere P, Nobile R, Panella FW, Squillace A (2006) Mechanical and microstructural behaviour of 2024–7075 aluminium alloy sheets joined by friction stir welding. Int J Mach Tools Manuf 46(6):588–594. https://doi.org/10.1016/j.ijmachtools.2005.07.010
10. Guo Y, Ma Ye, Zhang X, Qian X, Li J (2020) Study on residual stress distribution of 2024-T3 and 7075-T6 aluminum dissimilar friction stir welded joints. Eng Fail Anal 118:104911. https://doi.org/10.1016/j.engfailanal.2020.104911
11. Bussu G, Irving PE (2003) The role of residual stress and heat affected zone properties on fatigue crack propagation in friction stir welded 2024-T351 aluminium joints. Int J Fatigue 25(1):77–88. https://doi.org/10.1016/S0142-1123(02)00038-5
12. Sutton MA, Yang B, Reynolds AP, Taylor R (2002) Microstructural studies of friction stir welds in 2024-T3 aluminum. Mater Sci Eng A 323(1):160–166. https://doi.org/10.1016/S0921-5093(01)01358-2

13. Hamilton C, Dymek S, Kopyscianski M, Weglowska A, Pietras A (2018) Numerically based phase transformation maps for dissimilar aluminum alloys joined by friction stir-welding. Metals 8(5):324. https://doi.org/10.3390/met8050324
14. Hamilton C et al (2017) Application of positron lifetime annihilation spectroscopy for characterization of friction stir welded dissimilar aluminum alloys. Mater Charact 132:431–436. https://doi.org/10.1016/j.matchar.2017.09.017
15. Hatamleh O, Lyons J, Forman R (2007) Laser and shot peening effects on fatigue crack growth in friction stir welded 7075-T7351 aluminum alloy joints. Int J Fatigue 29(3):421–434. https://doi.org/10.1016/j.ijfatigue.2006.05.007
16. Hatamleh O, Singh PM, Garmestani H (2009) Corrosion susceptibility of peened friction stir welded 7075 aluminum alloy joints. Corros Sci 51(1):135–143. https://doi.org/10.1016/j.corsci.2008.09.031
17. Raja S, Manikumar R, Benruben R, Ragunathan S (2020) Effect of backing plate on strength and microstructural characteristics of friction stir welded AA2014-T6 aluminium alloy joints. Mater Today Proc. https://doi.org/10.1016/j.matpr.2020.02.938.
18. Bachmann A, Krutzlinger M, Zaeh MF (2018) Influence of the welding temperature and the welding speed on the mechanical properties of friction stir welds in en AW-2219-T87. In: 20th Chemnitz Seminar on Materials Engineering, WTK 2018, Chemnitz, Germany, vol 373, Institute of Physics Publishing, in IOP Conference Series: Materials Science and Engineering, 1 ed., https://doi.org/10.1088/1757-899X/373/1/012016
19. Magalhães A, Backer JD, Bolmsjö G (2019) Thermal dissipation effect on temperature-controlled friction stir welding. Soldagem & Inspeção 24. https://doi.org/10.1590/0104-9224/si24.28
20. Krutzlinger M (2018) Temperature control for Friction Stir Welding of dissimilar metal joints and influence on the joint properties
21. Pew JW, Nelson TW, Sorensen CD (2007) Torque based weld power model for friction stir welding. Sci Technol Weld Joining 12(4):341–347. https://doi.org/10.1179/174329307X197601
22. Cederqvist L, Reynolds AP, Sorensen CD, Garpinger O (2010) Reliable FSW of copper canisters using improved process and regulator controlling power input and tool Temperature. In: 8th International Symposium on Friction Stir Welding
23. Longhurst WR, Cox CD, Gibson BT, Cook GE, Strauss AM, DeLapp DR (2014) Applied torque control of friction stir welding using motor current as feedback. Proc Inst Mech Eng B: J Eng Manuf 228(8):947–958. https://doi.org/10.1177/0954405413514400
24. Davis TA, Ngo PD, Shin YC (2012) Multi-level fuzzy control of friction stir welding power. Int J Adv Manuf Technol 59(5–8):559–567. https://doi.org/10.1007/s00170-011-3522-0
25. Taysom BS, Sorensen CD, Hedengren JD (2016) Dynamic modeling of friction stir welding for model predictive control. J Manuf Processes 23:165–174. https://doi.org/10.1016/j.jmapro.2016.06.004
26. Babb JA, Steel R, Packer SM, Williams J (2010) Method for using modifiable tool control parameters to control the temperature of the tool during friction stir welding, US [Online]. Available: https://patents.google.com/patent/US20110172802A1/en
27. Ross K, Sorensen CD (2011) Investigation of methods to control friction stir weld power with spindle speed changes. In: TMS 2011
28. Das RRV, Kalaichelvi V, Karthikeyan R (2013) Application of fuzzy logic control strategy for temperature control in friction stir welding. In: ASME 2013 Gas Turbine India Conference, GTINDIA 2013. International Gas Turbine Institute, Bangalore. American Society of Mechanical Engineers (ASME). https://doi.org/10.1115/GTINDIA2013-3790.[Online]. Available:https://doi.org/10.1115/GTINDIA2013-3790
29. De Backer J, Bolmsjö G, Christiansson AK (2013) Temperature control of robotic friction stir welding using the thermoelectric effect. Int J Adv Manuf Technol 70(1–4):375–383. https://doi.org/10.1007/s00170-013-5279-0
30. Fehrenbacher A, Duffie NA, Ferrier NJ, Pfefferkorn FE, Zinn MR (2013) Effects of tool–workpiece interface temperature on weld quality and quality improvements through temperature

control in friction stir welding. Int J Adv Manuf Technol 71(1–4):165–179. https://doi.org/10.1007/s00170-013-5364-4
31. Bachmann A, Gamper J, Krutzlinger M, Zens A, Zaeh MF (2017) Adaptive model-based temperature control in friction stir welding. Int J Adv Manuf Technol 93(1–4):1157–1171. https://doi.org/10.1007/s00170-017-0594-5
32. Taysom BS, Sorensen CD (2020) Adaptive relay autotuning under static and non-static disturbances with application to friction stir welding, ISA Trans 97:474–484. https://doi.org/10.1016/j.isatra.2019.08.014
33. Taysom BS, Sorensen CD, Hedengren JD (2017) A comparison of model predictive control and PID temperature control in friction stir welding. J Manuf Processes 29:232–241. https://doi.org/10.1016/j.jmapro.2017.07.015
34. MatWeb (2020) Aluminum 2024-T4; 2024-T351. http://www.matweb.com/search/DataSheet.aspx?MatGUID=67d8cd7c00a04ba29b618484f7ff7524
35. MatWeb (2020) Aluminum 7075-T76; 7075-T7651. https://www.matweb.com/search/DataSheet.aspx?MatGUID=4f19a42be94546b686bbf43f79c51b7d
36. Nielsen BK (2009) Developing response surfaces based on tool geometry for a convex scrolled shoulder step spiral (CS4) friction stir processing tool used to weld AL 7075. MS, Mechanical Engineering, Brigham Young University. [Online]. Available: https://scholarsarchive.byu.edu/etd/1782/

Transitioning FSW to a Controlled Production Process

Arnold Wright, Devry Smith, Brandon Taysom, and Yuri Hovanski

Abstract Systematic investigation of the Friction Stir Welding (FSW) process shows that a fixed rotational velocity and feed rate may not yield uniform mechanical properties along the length of a weldment. Nevertheless, correlations between process parameters and post-weld material properties have successfully demonstrated that peak temperature and cooling rate drive post-weld properties. We review the reported methodologies reported for controlling friction stir welding with a detailed look at how temperature control has been used. We compare data from uncontrolled FSW of AA 6111-T4 sheet with controlled FSW at temperatures ranging from 375 °C to 450 °C, as a means of demonstrating that a simplified methodology of a single-loop PID controlling with spindle speed may be used to effectively control temperature. This methodology can be simply used with any machine that already has the ability to actively control spindle speed, and has been previously shown to be able to be auto-tuned with a single weld.

Keywords Temperature control · FSW · AA 6111

Introduction

Overview of Friction Stir Welding

The idea of using the frictional heat of rubbing metal parts together to weld was first patented in Britain in 1940, but did not see much interest until 1950, when the Soviet Union started research [1]. The process was quickly picked up by the Society of Automotive Engineering, where they published several papers related to their application in manufacturing automotive parts such as crankshafts. The initial welding machines were modified lathes, and power usage of 10% of the conventional

A. Wright (✉) · D. Smith · Y. Hovanski
Department of Manufacturing Engineering, Brigham Young University, Provo, UT, USA
e-mail: arnoldDwright94@gmail.com

B. Taysom
Pacific Northwest National Laboratory, Richland, WA, USA

Y. Hovanski et al. (eds.), *Friction Stir Welding and Processing XI*, The Minerals, Metals & Materials Series, https://doi.org/10.1007/978-3-030-65265-4_9

processes, along with cycle times under a minute were noted [2]. This process was later termed Rotary Friction Welding (RFW), and required at least one of the parts being joined to be cylindrical [3]. It took until 1991 for the idea of using a non-consumable tool to join arbitrary shapes to emerge, and this was called Friction Stir Welding [4]. In this process, the parts to be joined are fixtured in the desired configuration, and a spinning, non-consumable tool with a pin is plunged into the joint line. The tool then traverses along the seam, plastically deforming and melding the parts together without melting either one.

Why Control?

As Friction Stir Welding (FSW) is a friction-based process, the dependence of frictional heat generation on temperature and force end up creating a feedback loop [5]. This generally leads to the existence of a stable equilibrium temperature for a given feedrate and spindle power [6]. However, this stable equilibrium can require a significant amount of time to reach, requiring slow feedrates, and is also only applicable for some applications of FSW–welding large sheets provides a relatively stable environment, but the welding of small parts, cylinders, corner areas, or engine blocks does not [7–9]. Attempting to run constant-parameter welds in these applications can result in broken tools, excessive flash, or crashed heads [7–9].

Some may ask if lack of control has hindered commercial acceptance of the FSW process in industry. It has not. As early as 1995, a Scandinavian company was using FSW to assemble partial extrusions; in 1999, Boeing began using FSW to join sections of the Delta II and Delta IV rocket fuel tanks, and in 2004 the first commercial cargo airplane part entered into production for the C-17 aircraft [10]. However, constructing the proper weld parameter profile for these welds is not a simple task, and usually involves conducting several test welds and destructive testing, even with an experienced operator. Conversely, installing and implementing temperature control on an existing FSW machine can be an expensive and labor-intensive process, requiring the installation and tuning of various motors, sensors, and controllers. Any control method desiring to be implemented in industry must require little instrumentation or have a simple, cost-effective integration to have widespread acceptance.

Why Temperature Control?

As every aspect of FSW interacts with almost every other, it is difficult to reliably control one measurement without also affecting the others [5]. Conversely, if you can reliably control one parameter of the weld, you can increase the overall weld quality and repeatability, with some parameters having a stronger link to quality than others [11]. Temperature has reliably been shown to have a strong link to post-weld properties such as fracture toughness, corrosion resistance, and grain size (strength), particularly with heat-treatable alloys [6, 8, 12]. Controlling temperature outside of the natural equilibrium can also improve tool life, and reduce defects such as wormholes [6, 7]. Temperature control can also start with very high rpm to achieve initial temperature and properties at the beginning of a weld without later running into thermal runaway.

Temperature Control in FSW

While FSW was invented in 1991, methods for any type of closed-loop control did not appear until around 2003 [13]. Investigation of temperature control specifically was spearheaded by the Cederqvist, Fehrenbacher, and Sorensen research groups starting around 2009–10 [14–16]. A decade later, and 30 years after the invention of FSW, we can review the major methods of temperature control. We focus on the control method itself, although there are also differences in the method of measuring the temperature itself, with the main methods being an embedded thermocouple [16] or through the tool-workpiece-thermocouple method [12].

Open-loop and Manual Control

Open-loop is the primeval control method for FSW. Experimentation through trial-and-error finds the weld parameters such as spindle speed, travel speed, and spindle downforce required to make a weld that satisfies the customers' strength requirements. Finding these parameters can be an expensive, lengthy process, even if experienced consultants are brought in. Once these parameters are found, the machine is programmed and these parameters are never changed unless defective welds are discovered. Programs where the parameters are constant [6], or where they are adjusted in steps at pre-determined weld positions, fall under open-loop control.

Manual is the most basic level of temperature control, where an operator adjusts parameters such as spindle speed, feedrate, or tool downward force [7] on-the-fly in order to maintain satisfactory weld temperatures. A skilled operator with experience with the specific operation can maintain some welds within 30 °C of the temperature setpoint [17]; however, achieving this level of skill can take many practice welds, and errors can still occur. Most methods of computerized control can achieve a more accurate and repeatable weld.

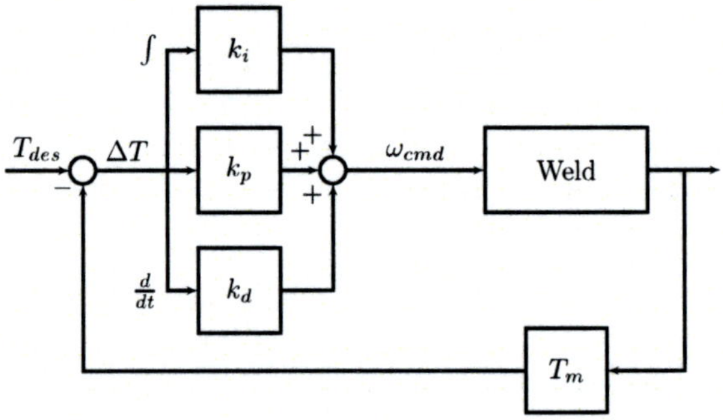

Fig. 1 Control block diagram for a PID temperature controller based on spindle speed

PID Spindle Speed

Single-loop Proportional-Integral-Derivative (PID) control of temperature is a fixture in many other areas of industry, and has also been implemented in FSW. Common variants of this controller scheme are P, PI, and PD where one or more of the gains are set to 0. Tuning can be accomplished by the Ziegler-Nichols method, which involves manual fine-tuning over repeated welds, or through other tuning rules. Noteworthy for this control method, an effective auto-tuner has also been designed [18], and in an implementation on the authors' machine results in an almost plug-and-play situation [19]. A version of this controller with the derivative gain set to 0 was investigated by Fehrenbacher in relation to control of welding in aluminum [11]. The removal of the derivative term was due to the sinusoidal changes in the measured temperature on an exposed shoulder thermocouple. This introduced a magnified sinusoidal component into the commanded spindle speed, and was detrimental to the performance of the controller. Fehrenbacher's tuning was accomplished through controller design, rather than manual adjustment. The controller was able to successfully reduce the temperature error due to disturbances to within 10 °C, as compared to within 40 °C uncontrolled [11]. This control method is a popular first step in implementing temperature control on a new system, due to its ease of integration into an existing machine, requiring only temperature measurement and the ability to command spindle speed [20, 21] (Fig. 1).

PID Feedforward

This method was first proposed in FSW by Cederqvist for use in the sealing of copper canisters [14]. It combines a simple Proportional-Derivative (PD) controller, with the spindle speed being adjusted based on the tool temperature, and an Integral (I) controller based on the weld power to correct for variations between welds. Weld

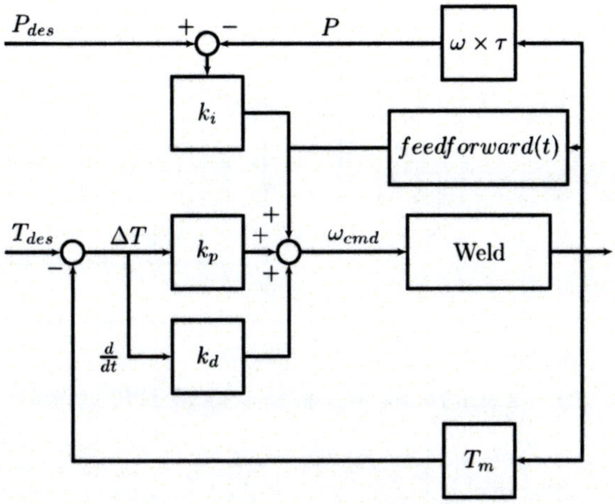

Fig. 2 Control block diagram for a PID temperature controller with an additional input for desired power and a feedforward term

power is determined through the equation: P = ω × τ. The overall Proportional-Integral-Derivative (PID) gains could be calculated with a variation on the Ziegler-Nichols method called AMIGO [14]. The feedforward term, however, requires data from previous successful welds, and Cederqvist also found that power requirements could vary substantially from one weld to the next [14]. This led to the idea of using a cascade controller, which could determine the weld power requirement on-the-fly [7]. However, all methods that use weld power as a controlling variable require measuring not only the spindle speed, but also the spindle torque. Measurement of the welding torque can be difficult, requiring a torque-calibrated motor or other significant adjustments to the welding machine (Fig. 2).

PID Cascade—Spindle Speed

This is an improved version of the PID-Feedforward method, where the variations between different welds can be self-corrected by the controller in the outer loop. This results in a much more accurate and reliable controller which requires less operator management. The inner loop monitors the weld power, and adjusts the spindle speed to keep the weld input power constant. The weld power is used because it is closely linked to the weld temperature, but was a faster indicator with regards to disturbances than the available temperature measurement system [22]. The outer loop adjusts the desired weld input power to keep the weld temperature constant. As proposed, the controllers for this method can be either PI or PID in form, as FSW is not a very dynamic process [7]. The tuning of the controller was carried out with a method that corrected for the noise in the measured signals in calculating the derivative gain [23]. For the tuning, two stages of step response experiments need to be performed, and manual fine-tuning was required before the system was considered reliable [7]. Once properly programmed, this method was producing

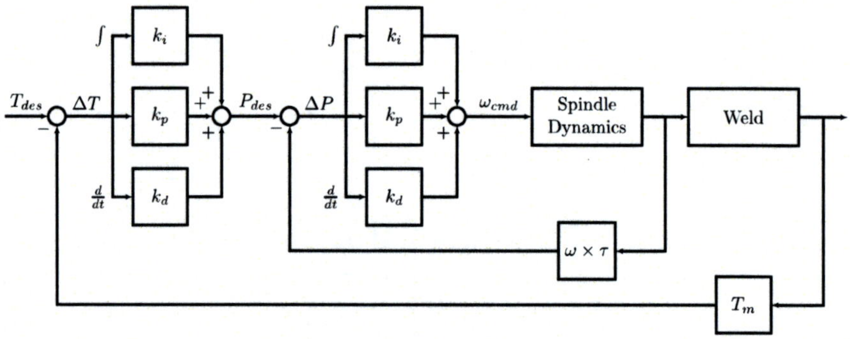

Fig. 3 Control block diagram for a cascaded PID temperature controller

welds in copper with a control error standard deviation of 1.6 °C, with a maximum deviation of 13.1 °C [7], which was a great improvement over manual control. This method was recommended by both Cederqvist [7], and Mayfield and Sorensen [6] in 2010. A potential improvement by Cederqvist used a pre-set weld start procedure to improve the initial rise to temperature [24] (Fig. 3).

PID Cascade—Torque

A variation on the PID Cascade control method was introduced by Ross and Sorensen [25]. Instead of controlling the spindle speed in the inner loop, the controlled parameter is the spindle torque, and only the spindle speed is used as the feedback signal. This improves the resulting control, as the disturbance response with torque is more stable than spindle speed, and, additionally, spindle speed has a finer resolution and is generally a better feedback signal. After manual tuning, experimental welds in aluminum and steel had standard deviations in temperature of 0.72 °C and 1.27 °C, respectively [26] (Fig. 4).

Model-based Control

Model-based control is a relatively new method of temperature control. Rather than PID gains being tuned, this method uses a process model to relate manipulated variables such as spindle speed and feedrate with controlled variables such as temperature [9]. A review of two types of model-based controllers found that they did not perform appreciably better than well-tuned PID controllers in a control scenario with ~1.5 s of time delay in the tool temperature [27]. Model-based controllers have advantages for systems that are complex, highly non-linear, have a large time delay to time constant ratio, or have modelable disturbances, none of which characterize

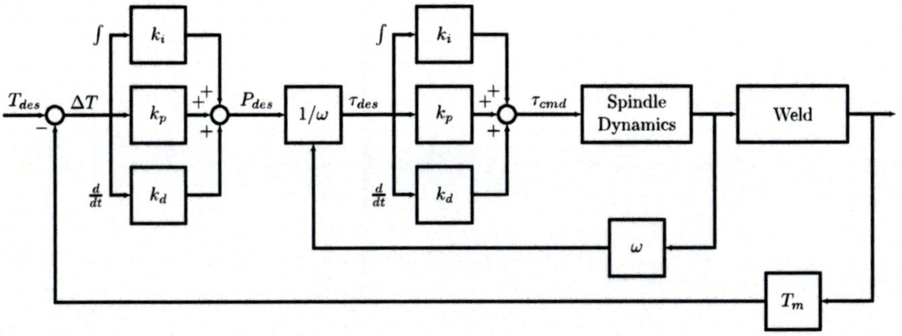

Fig. 4 Control block diagram for a cascaded PID temperature controller, where the inner loop controls the weld torque

typical FSW [27]. Consequently, PID controllers usually have similar performance to model-based controllers for FSW in most cases.

Dual Temperature and Force Control

While we focus on temperature control, it is worth mentioning that there have also been approaches to combine temperature and force control. One approach is a linked MIMO system to reject disturbances to both temperature and plunge depth [28]. However, initial results are not significantly different from pure temperature control. Others have simply engaged a depth controller at the same time as temperature control, and this resulted in an overall better-formed weld than temperature control alone [29].

Methods

After evaluating the different temperature control methods for ease of implementation and use, the simplicity of the single-loop PID controller, combined with the auto-tuner, lead to its selection for evaluation. Temperature measurement was performed with a 0.8 mm diameter type K thermocouple (Omega SCASS-032U-12) embedded in a H-13 tool of 12 mm shoulder diameter. The thermocouple reached to within 0.5 mm of the shoulder, as shown in Fig. 5. The signal from the thermocouple was transmitted using a TC-Link -1CH -LXRS node attached to the spindle, which wirelessly transmitted over Bluetooth to a WSDA -BASE -101 -LXRS analog base station. The base station logged the temperature, and then transmitted a remapped thermocouple voltage on a 0–3.3 V scale to the PLC. Temperature was filtered with a sum of squared error (SSE) filter [30] in order to achieve smooth temperature and first-derivatives for temperature control.

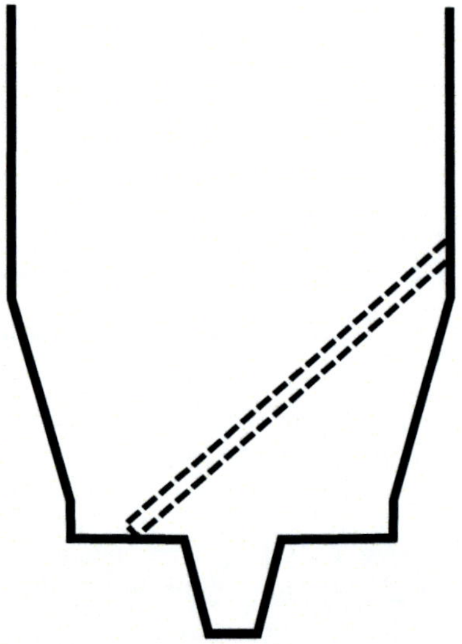

Fig. 5 Cross-section of the tool showing the thermocouple position in relation to the shoulder

To determine the gains for PID control, an adaptive relay test using a time based correction [19] was used to determine approximate system parameters in a bead-on-plate weld at the desired feedrate. Preliminary PID tuning gains were then calculated by taking the logarithmic mean of servo and regulator gains [27, 31]. Instead of using these default gains, the present work uses PID gain scheduling to avoid integrator windup and temperature overshoot in the beginning of the weld for better performance. PD gains were used from before the plunge until the temperature was within 5 °C of the setpoint. Once the temperature was within 5 °C of the setpoint, the integral gain was changed from 0 to the calculated value, and used until the end of the weld. This anti-integrator windup measure lowered overshoot on the roughly 350 °C step input in temperature at the beginning of the weld.

To test the capabilities of the auto-tuner and controller over a wide range of conditions, a parameter grid of possible welds was laid out. Two feedrates were chosen: 1000 mm/min and 2000 mm/min; and four temperature control setpoints of 375, 400, 425, and 450 °C, with an additional weld for each feedrate with no control. For each set of conditions, one auto-tuning weld was performed, and then one weld was run using the calculated PID settings. After this weld, the machine was allowed to cool to approximately 30 °C before the next tuning weld was performed. Welding was performed on a TTI RM-2 type linear Friction Stir Welding machine retrofitted with a Bond Technologies high-speed B&R based controller. The base material was AA6111-T4, and each coupon was 200 × 1000 × 2.7 mm in size, resulting in a welded specimen size of 400 × 1000 × 2.7 mm. Welds were 900 mm in length, and the weld and machine parameters were logged at a rate of 1250 Hz. A 400 × 100 × 6 mm plate of mild steel was placed under the weld to protect the main anvil of 50 mm thick steel. Two 12 × 25 mm bars of mild steel were used to evenly distribute the clamping load from four step clamps along the length of the coupons. Four screw

Fig. 6 Weld setup

blocks were used to provide a lateral clamping force. An image of the overall setup is given in Fig. 6.

Results

As presented in Fig. 7, the temperature-controlled welds had a slower initial rise time as compared to the weld with no control, but quickly stabilized at the control setpoint

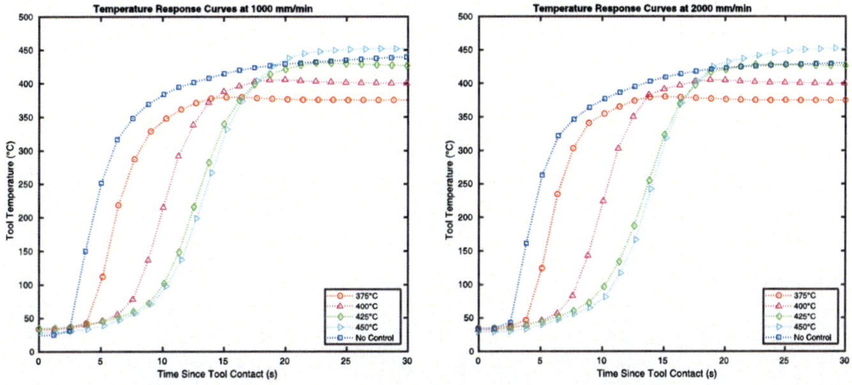

Fig. 7 Weld temperature response curves by feedrate and time. (Color figure online)

with minimal overshoot. As temperature control was enabled before the plunge, a longer dwell time could be safely managed without risk of melting. This allowed a greater portion of the weld to be completed at the desired process temperature. This distinction between the time domain and the weld position domain is made more clear in Fig. 8, where the temperature response is plotted according to the tool position.

The percent overshoot, 10–90% rise time, 5 °C settling time, and standard deviation of temperature after settling are included in Table 1. The value of 5 °C for the settling time was chosen as a reasonable approximation of a 2% settling time, as the 10% settling times listed in other papers do not have much meaning when the maximum overshoot is 1–2%.

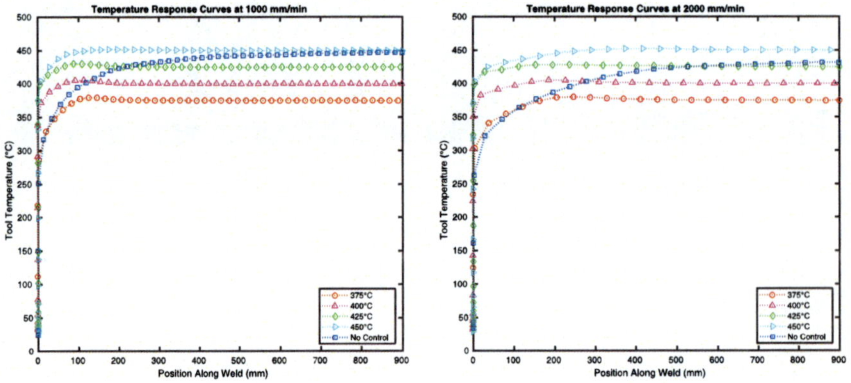

Fig. 8 Weld temperature response curves by feedrate and weld position. (Color figure online)

Table 1 Results of temperature control

Feedrate (mm/min)	Setpoint temperature	Overshoot (%)	Overshoot (°C)	Rise time (s)	Settling time (s)	Std. dev. (°C)
1000	375 °C	1.53	5.2	5.10	16.0	1.00
	400 °C	1.70	6.2	6.15	20.8	0.92
	425 °C	1.52	5.9	7.77	24.3	1.10
	450 °C	0.46	1.9	8.91	21.6	0.87
2000	375 °C	1.69	5.7	4.43	16.0	1.25
	400 °C	1.51	5.6	5.87	19.9	1.18
	425 °C	0.83	3.3	7.96	19.9	1.30
	450 °C	0.62	2.6	8.35	24.0	1.36

Discussion

The auto-tuned controller was able to provide fast and stable control at a range of temperatures, with resulting rise times, settling times, and standard deviations within the ranges reported by other authors for various methods of control, as shown in Fig. 9. The temperature response rises quickly enough that only the first 100 mm for 1000 mm/min welds and the first 200 mm for 2000 mm/min welds see temperatures more than 5 °C away from the setpoint temperature. This difference is due primarily to the feedrate, as the settling times are similar. Compare this to the uncontrolled weld, where the temperature is continually increasing, with the start and end portions of the weld having been welded at a nearly 100 °C difference in temperature. Starting the temperature controller before the plunge allows the weld temperature to be brought closer to the desired temperature before the traverse, without the risk of a large temperature overshoot later.

A decade ago, when temperature control was in its infancy, and often required specialty motors or expensive hardware, it may have been difficult to justify the expense of adding control against the cost of manual control. However, in today's world with inexpensive and fast temperature measurement, there is little reason to not add temperature control to production machinery. Tuning for a new part can be easily accomplished by sacrificing one or two parts at the beginning of a production run. Improving the consistency of the conditions experienced by the weld nugget should improve consistency in material properties along the length of the weld, thus improving overall part quality. Implementing temperature control should also reduce inter-batch and intra-batch consistency, as it can correct for changing environmental

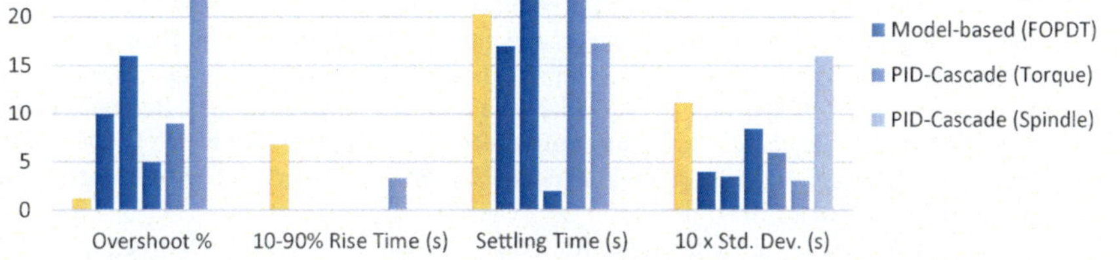

Fig. 9 Comparison of the results from using an autotuned PID controller based on spindle speed with results reported by several other authors. Rise times were not available from all sources. The PID-spindle "Regulator" and "Servo", and the two model-based results are in AA7075 from [27], the PID-cascade-torque results are in AA7075 from [26], and the PID-cascade-spindle results are in copper from [7]. (Color figure online)

conditions. Temperature control also improves the ability of FSW joins to be made in arbitrary shapes, rather than just focusing on joining sheet-like workpieces.

Conclusions

With 30 years of friction stir welding and a decade of experience in thermal control of the process we see surprisingly low implementation of thermal control. Numerous authors have reported on the ability to accurately control the process leading to improvements in weld properties, reduction in development time, and greater ability to deal with process disturbances. While several temperature control methodologies require complex adaptations of motor drivers and controllers, we demonstrated that using a *simplified* single-loop PID control of spindle speed may be used to effectively control temperature. Furthermore, we showed this approach can be used in conjunction with an auto-tuning setup to allow essentially push-button operator use of thermal control. This simplified setup can be easily adapted to any machine that already allows for digital spindle speed control.

References

1. Ellis CR (1970) Friction welding: what it is and how it works. Met Constr Br Welding J 2(5):185–188
2. Chyle JJ (1960) Prediction in new metal joining processes. SAE Trans 68:143–156
3. Hollander MB, Cheng CJ, Quimby JA (1964) Friction welding-A "Natural" for production. SAE Trans 73:242–254
4. Anon (1991) Linear friction welding joins noncircular sections. Adv Mater Process 139(2):47
5. Colligan KJ, Mishra RS (2008) A conceptual model for the process variables related to heat generation in friction stir welding of aluminum. Scr Mater 58(5):327–331. https://doi.org/10.1016/j.scriptamat.2007.10.015
6. Mayfield D, Sorensen CD (2010) An improved temperature control algorithm for friction stir processing. In: Paper presented at 8th International symposium on friction stir welding, Timmendorfer Strand, Germany
7. Cederqvist L, Garpinger O, Hagglund T, Robertsson A (2010) Cascaded control of power input and welding temperature during sealing of spent nuclear fuel canisters. In: Paper presented at ASME Dynamic Systems and Control Conference (DSCC), Cambridge, Massachusetts, USA
8. Silva-Magalhães A, Cederqvist L, De Backer J, Håkansson E, Ossiansson B, Bolmsjö G (2019) A friction stir welding case study using temperature controlled robotics with a HPDC cylinder block and dissimilar materials joining. J Manuf Process 46:177–184. https://doi.org/10.1016/j.jmapro.2019.08.012
9. Bachmann A, Gamper J, Krutzlinger M, Zens A, Zaeh MF (2017) Adaptive model-based temperature control in friction stir welding. Int J Adv Manuf Tech 93(1–4):1157–1171. https://doi.org/10.1007/s00170-017-0594-5
10. Kallee SW (2010) Industrial applications of friction stir welding. In: Lohwasser D, Chen Z (eds) Friction stir welding: from basics to applications. Woodhead Publishing Limited, 118–163
11. Fehrenbacher A, Duffie NA, Ferrier NJ, Pfefferkorn FE, Zinn MR (2011) Toward automation of friction stir welding through temperature measurement and closed-loop control. J Manuf Sci Eng 133(5). https://doi.org/10.1115/1.4005034

12. Fehrenbacher A, Schmale JR, Zinn MR, Pfefferkorn FE (2012) Tool-workpiece interface temperature measurement in friction stir welding. Paper presented at ASME International Manufacturing Science and Engineering Conference (MSEC), Notre Dame, IN, USA
13. Cook GE, Smartt HB, Mitchell JE, Strauss AM, Crawford R (2003) Controlling robotic friction stir welding. Weld J 82(6):28–34
14. Cederqvist L, Johansson R, Robertsson A, Bolmsjo G (2009) Faster temperature response and repeatable power input to aid automatic control of friction stir welded copper canisters. Paper presented at friction stir welding and processing V-TMS annual meeting & exhibition, San Francisco, CA, USA
15. Fehrenbacher A, Duffie NA, Ferrier NJ, Zinn MR, Pfefferkorn FE (2010) Temperature measurement and closed-loop control in friction stir welding. Paper presented at 8th International symposium on friction stir welding, Timmendorfer Strand, Germany
16. Cederqvist L, Reynolds AP, Sorensen CD, Garpinger O (2010) Reliable FSW of copper canisters using improved process and regulator controlling power input and tool temperature. Paper presented at 8th International symposium on friction stir welding, Timmendorfer Strand, Germany
17. Cederqvist L, Garpinger O, Hagglund T, Robertsson A (2011) Reliable sealing of copper canisters through cascaded control of power input and tool temperature. In: Mishra RS, Mahoney MW, Sato Y, Hovanski Y, Verma R (eds) Friction stir welding and proceesing VI 2011 TMS
18. Marshall D, Sorensen C (2013) System parameter identification for friction stir processing. Paper presented at Friction stir welding and processing VII-TMS annual meeting & exhibition, San Antonio, TX, USA
19. Taysom BS, Sorensen CD (2020) Adaptive relay autotuning under static and non-static disturbances with application to friction stir welding. ISA Trans 97:474–484. https://doi.org/10.1016/j.isatra.2019.08.014
20. De Backer J, Bolmsjö G, Christiansson A-K (2013) Temperature control of robotic friction stir welding using the thermoelectric effect. Int J Adv Manuf Tech 70(1–4):375–383. https://doi.org/10.1007/s00170-013-5279-0
21. Krutzlinger M (2018) Temperature Control for Friction Stir Welding of Dissimilar Metal Joints and Influence on the Joint Properties. Tribology in Manufacturing Processes and Joining by Plastic Deformation II 2018. Trans Tech Publications Ltd, 360–368
22. Ross K, Sorensen CD (2011) Investigation of Methods to Control Friction Stir Weld Power with Spindle Speed Changes. Paper presented at TMS 2011 Annual Meeting & Exhibition, San Diego, CA, USA
23. Garpinger O (2009) Design of Robust PID Controllers with Constrained Control Signal Activity. Licentiate Thesis, Lund University
24. Cederqvist L, Garpinger O, Hagglund T, Robertsson A (2012) Cascade control of the friction stir welding process to seal canisters for spent nuclear fuel. Control Eng Pract 20(1):35–48. https://doi.org/10.1016/j.conengprac.2011.08.009
25. Ross K, Sorensen CD (2013) Paradigm Shift in Control of the Spindle Axis. Paper presented at Friction Stir Welding and Proceesing VII - TMS Annual Meeting & Exhibition, San Antonio, TX, USA
26. Ross K, Sorensen CD (2013) Advances in Temperature Control for FSP. In: Mishra RS, Mahoney MW, Sato YS, Hovanski Y, Verma R (eds) Friction Stir Welding and Proceesing VII 2013. TMS
27. Taysom BS, Sorensen CD, Hedengren JD (2017) A comparison of model predictive control and PID temperature control in friction stir welding. J Manuf Process 29:232–241. https://doi.org/10.1016/j.jmapro.2017.07.015
28. Fehrenbacher A, Smith CB, Duffie NA, Ferrier NJ, Pfefferkorn FE, Zinn MR (2014) Combined temperature and force control for robotic friction stir welding. J Manuf Sci Eng 136(2). https://doi.org/10.1115/1.4025912
29. Cederqvist L, Garpinger O, Nielsen I (2017) Depth and Temperature Control During Friction Stir Welding of 5cm Thick Copper Canisters. Springer International Publishing, 249–260

30. Taysom BS, Sorensen CD (2019) Advances in Signal Processing for Friction Stir Welding Temperature Control. Friction Stir Welding and Processing X, 135–147
31. O'Dwyer A (2006) Handbook of PI and PID Controller Tuning Rules. Imperial College Press, London

Removing Rotational Variations from Shoulder Thermocouples in Friction Stir Welding

Brandon Scott Taysom, WoongJo Choi, and Kenneth Ross

Abstract Thick plate and lower temperature friction stir welding results in good post-weld properties by welding at very low tool rotational speeds. Low speeds cause large variations in the measured shoulder temperature, making temperature control difficult. This paper describes a computationally light method to correct for oscillating measured temperatures. Data is collected over the previous two revolutions and used to build a compensation table and calculate a derivative. In simulation, measured temperature variations are reduced by a factor of five to ten. Temperature data is preserved and not cornered, nor is any time delay induced by this method, unlike filtering methods.

Keywords Friction stir welding · Temperature · Temperature control

Introduction

Friction stir welding (FSW) is a solid-state welding process used to join sheets and plates with very high strength [1]. In FSW, a hardened tool is rotated and plunged into the seam of the workpieces. This creates heat, softening the workpieces, and allowing them to be mixed together and joined [2]. FSW has been used to successfully join thick plates (2 inches and over) of soft alloys such as copper [3, 4] and aluminum [5].

FSW is a thermomechanical process, and as such temperature is one of the most important process parameters [2, 6]. Temperature is neither constant nor uniform in and around the weld. Viewed from a transverse section, the inner portion of the weld, where the majority of the deformation occurs (the nugget), is generally the hottest part of the weld, with temperature decreasing away from the tool [7, 8]. Viewed from above, the advancing side of the weld is generally hotter than the retreating side [7, 9]. In addition, the temperatures change over time during a weld [10].

B. S. Taysom (✉) · W. Choi · K. Ross
Pacific Northwest National Laboratory, Richland, USA
e-mail: brandon.taysom@pnnl.gov

Y. Hovanski et al. (eds.), *Friction Stir Welding and Processing XI*, The Minerals, Metals & Materials Series, https://doi.org/10.1007/978-3-030-65265-4_10

Controlling the process temperature can lead to better post-weld properties [6, 11]. In order to control the temperature of a weld, a point of temperature measurement must be decided upon, usually in one of two locations in a tool [6]. Thermocouples are often inserted in the pin of a tool; this method works well for shorter pins and/or metal tools [12, 13]. Thermocouples are also commonly located in the shoulder of the tool [3, 14, 15]. This is done when it is not feasible to insert a thermocouple into the pin, when a measurement at the shoulder–workpiece interface is desired, and when the tool face is made of a material such as PCBN [14, 15]. In these cases, a shoulder thermocouple is often used when welding steels and thick-section copper and aluminum.

Shoulder thermocouple measurements include a cyclic temperature component, making precise temperature control difficult. In the steady-state portion of a weld, a thermocouple in the shoulder goes over both the advancing and retreating sides of the weld. Due to the different temperatures of the advancing and retreating sides [7, 9], a fixed point in the tool shoulder experiences a varying heat flux, leading to a fluctuating measurement in a process that is otherwise steady state.

Previously, a method that has been developed to compensate for a cyclic offset create by a radio telemetry collar in FSW [16]. In this case, the offset was only a function of angle, and could thus be modeled before a weld and removed during the welding process. In contrast, the amount of variation in the advancing and retreating side is mostly proportional to the temperature, meaning that the offset cannot be accounted for in advance. In addition, process changes can affect the temperature differences between the advancing and retreating sides of the weld, necessitating a dynamically updating compensation method.

This paper describes a method that automatically subtracts the rotational component of measured temperature while preserving the non-rotational components. This is done in a way that has an extremely lightweight PLC footprint, can operate without needing any user input, and unlike many filters adds no time delay.

Correcting Rotational Temperature Effects

During steady-state welding, the shoulder temperature varies in a periodic fashion as the thermocouple rotates between the advancing and retreating side of the weld. The measured temperature may or may not be nearly sinusoidal, and thus relying on a sinusoidal fit to compensate for the observed changes may be insufficient. Instead of a single sinusoidal fit, a spline is used to capture the rotational variations as this can accommodate more complex waveforms.

For the present work, a cubic spline is used to capture the temperature variations within a single revolution, with end-point extrapolation being performed linearly. Temperature data is collected over a revolution and binned into 12 equal segments from 0 to 30°, 30 to 60°, etc. The temperature data in each segment is averaged, and the mean of the data is subtracted from each point individually to give basis for the spline compensation. This is shown in simulation for a sinusoidal temperature deviation in

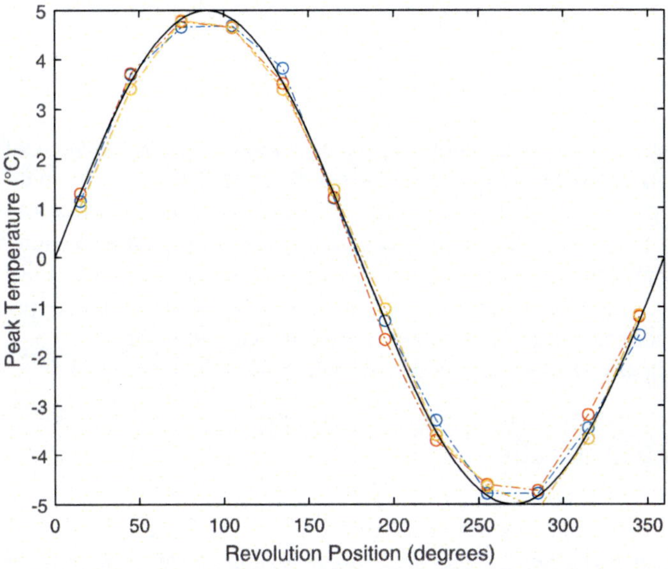

Fig. 1 Simulated temperature deviation compared to compensation spline basis points. The black line represents the "actual" temperature deviation from the steady-state temperature. Noise of 0.5 °C was added to the actual temperature, and was collected, binned, and averaged for each revolution. The basis points are shown by circles, and connected with dashed lines to aid visual recognition. Three separate revolutions are shown. (Color figure online)

Fig. 1. Overall, the waveform is captured well by the basis points, although some noise-based variation occurs, as well as clipping of the peaks.

For this method, data is collected during revolution *n-1* and *n-2*, and the averaged data is used to build a correction spline for revolution *n*. A cubic spline is used on the interior of the basis points (between 15° and 345° based upon a spacing of 30°), with linear extrapolation used outside the domain of the basis points.

This method captures trends well during steady state, but falls short when the average temperature is not constant. If the temperature is increasing, then the basis points will have a similar upward trend. If applied in this way, a stair-step artifact will occur in the resulting compensated data, which is physically unrealistic. Instead, a derivative is calculated from the *n-1* and *n-2* revolutions, and the derivative is subtracted from each revolution's basis points. Thus, when the basis is applied, only the cyclic deviation is removed, leaving the underlying trends in place. If the process is at steady state, then this derivative correction has no additional effect.

The robustness of the process is increased in two ways. If the rotational speed of the spindle is fast relative to the data collection rate, then it is possible that no data is binned into a segment, leading to a potential 0/0 situation when taking the average of each segment's data. When this occurs, the segment can be set to either the average of all other measurements, or the average of its non-zero neighbors. An accuracy penalty is thus incurred to avoid ∞, NaN, or 0 results. Another error can occur if the spindle is parked. In this case, temperature data is continually added to a single segment, which can eventually lead to numerical overflow. Adding an arbitrarily high cap to the number of data points recorded per segment (for instance, 10,000) will circumvent this issue.

Because this method continually recalibrates every revolution, small errors are quickly expelled from the system. A good example of this is in spindle position. An encoder will keep spindle position in sync even after thousands of revolutions, whereas using rotational speed to continually estimate spindle position will eventually accumulate error. Since this method only looks backward two revolutions, any error resulting from estimating angular position based upon spindle speed will only have a moderate effect, and cannot compound with future error.

Simulation Results

Two sets of simulations were performed to gauge the effectiveness of this method. Simulations use a data collection rate about 50 times faster than the rotational speed, resulting in about four measurement points per spline basis point. First, the behavior at steady state with a simple sinusoidal temperature deviation was investigated. Next, the behavior under a more complex deviation for both transient and steady-state operation was investigated. Finally, a closed-loop simulation was performed with PID control based upon the corrected shoulder temperature.

For the first simulation, a steady-state process with sinusoidal thermocouple deviation of 5 °C with 0.5 °C noise at 150 rpm was simulated. The temperature errors were calculated against the underlying baseline temperature without deviation or noise. A plot of the error over time is shown in Fig. 2. The average absolute error decreases by over ten times—from 3.2 °C to 0.28 °C.

For the second simulation, a more complex temperature deviation profile was imposed, and a steep rise in temperature followed by a sharp leveling off was simulated. For this simulation, the nominal oscillation amplitude was 10 °C, with a more aggressive noise of 2 °C. The temperature profiles are shown in Fig. 3, with the error shown in Fig. 4. During both the transient and steady-state portion of the simulation, the absolute average temperature error was about 5.7 °C and 1.1 °C for the measured and corrected temperatures, respectively.

Finally, a closed-loop PID simulation was performed to evaluate the performance of the correction method in a slightly more realistic manner. Noise was set at 1 °C, and the amplitude of temperature variation was set to be 1% of the nominal temperature. The system was modeled as a first-order plus dead time system with a time constant of 2 s, time delay of 0.3 s, gain of 5 W/°C, and PID parameters were chosen as $k_p = 3$ W/°C, $k_i = 1.5$ W/s°C, $k_d = 0$. The PID controller referenced the real-time compensated signal and did not know the "true" temperature. The response of the system to a step change from 725 °C to 800 °C is shown in Fig. 5. The temperature rise is stable and smooth, and the temperature measurement error is reduced by a factor of 9.

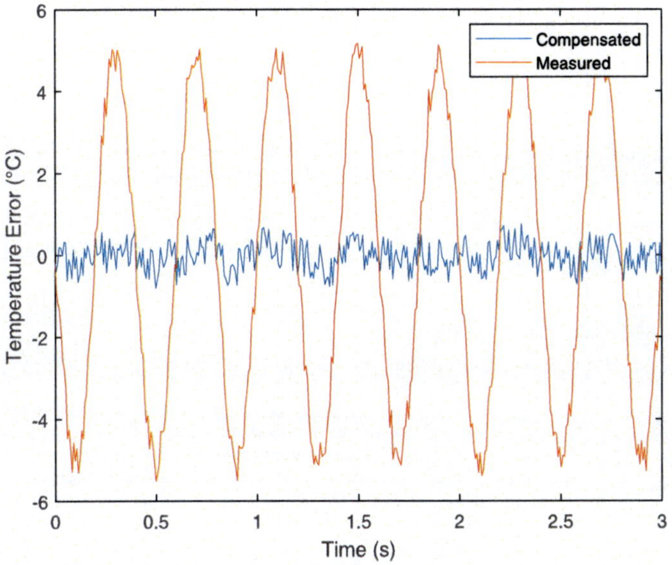

Fig. 2 Error of measured temperature compared to corrected temperature. Temperature error drops by over ten times, and the resultant temperature is mostly flat. (Color figure online)

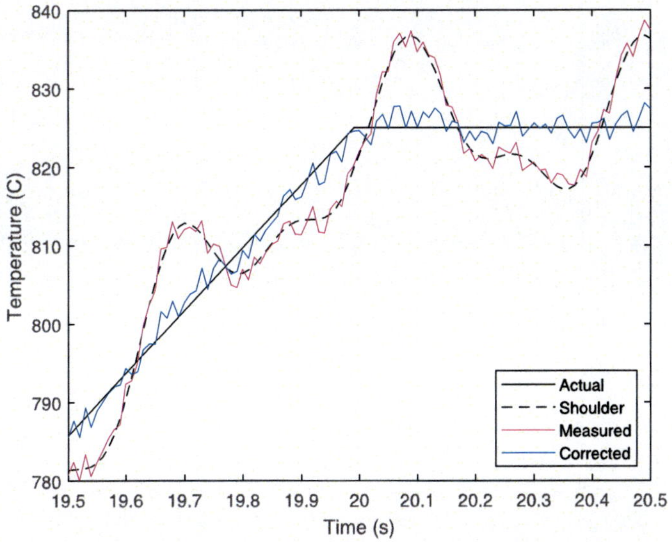

Fig. 3 Temperature profile during a transient to steady-state simulation. The underlying average temperature is shown in a black line, with the temperature at a rotating point shown by the black dashed line. The raw measured temperature (including measurement noise) is shown in magenta, with the corrected measured temperature in blue. Overall, the corrected temperature is very close to the baseline temperature, and slight filtering could reduce the observed noise if desired. (Color figure online)

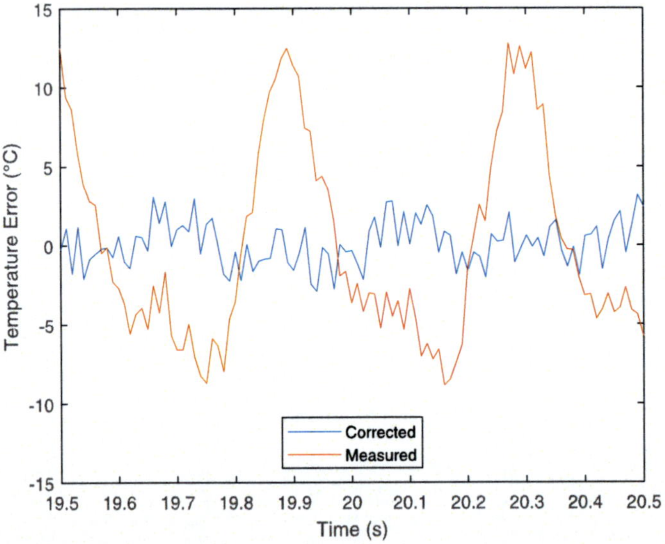

Fig. 4 Error profiles of the measured and corrected temperatures. The correction method reduces temperature deviation by a factor of approximately five. (Color figure online)

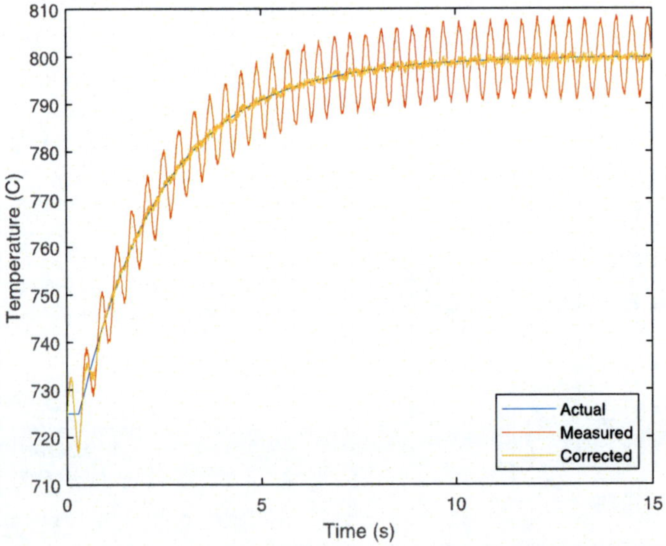

Fig. 5 Temperature response of system under PID control. (Color figure online)

Conclusion

A correction method was developed to remove oscillation in a shoulder thermocouple in an otherwise steady-state process. The correction method dynamically adapts to changes in the process, is computationally lightweight to enable implementation on a PLC, and requires no interaction by the end user.

In simulation, this method removed 80–90% of the shoulder temperature oscillation from a measured signal. The method adds zero time delay to the system, unlike an aggressive butterworth filter. This method can be used as the control variable for a PID controller resulting in in smooth temperature response. Future work includes implementing this on a FSW machine to test real-world performance.

References

1. Lakshminarayanan AK, Balasubramanian V, Elangovan K (2009) Effect of welding processes on tensile properties of AA6061 aluminium alloy joints. Int J Adv Manuf Technol 40(3–4):286–296
2. Mishra R, Ma Z (2005) Friction stir welding and processing. Mater Sci Eng R Rep 50(1–2):1–78
3. Cederqvist L, Garpinger O, Hägglund T (2001) Reliable Sealing of copper canisters through cascaded. Friction Stir Weld and Process VI:51–58
4. Cederqvist L, Garpinger O, Hägglund T, Robertsson A (2012) Cascade control of the friction stir welding process to seal canisters for spent nuclear fuel. Control Eng Pract 20(1):35–48
5. Prime MB, Gnäupel-Herold T, Baumann JA, Lederich RJ, Bowden DM, Sebring RJ (2006) Residual stress measurements in a thick, dissimilar aluminum alloy friction stir weld. Acta Mater 54(15):4013–4021
6. Fehrenbacher A, Duffie NA, Ferrier NJ, Pfefferkorn FE, Zinn MR (2014) Effects of tool-workpiece interface temperature on weld quality and quality improvements through temperature control in friction stir welding. Int J Adv Manuf Technol 71(1–4):165–179
7. Meenu SV, Misra JP (2017) Study on temperature distribution during Friction Stir Welding of 6082 aluminum alloy. Mater Today Proc 4(2):1350–1356
8. Tang W, Guo X, McClure JC, Murr LE, Nunes A (1998) Heat input and temperature distribution in friction stir welding. J Mater Process Manuf Sci 7(2):163–172
9. Maeda M, Liu H, Fujii H, Shibayanagi T (2005) Temperature field in the vicinity of FSW-tool during friction stir welding of aluminium alloys. Weld. World 49(3–4):69–75
10. Taysom BS (2015) Temperature control in friction stir welding using model predictive control. Brigham Young University
11. Sorger G, Sarikka T, Vilaça P, Santos TG (2018) Effect of processing temperatures on the properties of a high-strength steel welded by FSW. Weld World 62(6):1173–1185
12. Taysom BS, Sorensen CD, Hedengren JD (2016) Dynamic modeling of friction stir welding for model predictive control. J Manuf Process 23
13. Ross KA, Sorensen CD (2013) Advances in temperature control for FSP. Friction Stir Weld Process VII:301–310
14. Allan T, Nelson T (2015) Study on the Fracture Toughness of Friction Stir Welded API X80. Eng Fract Mech 150:58–59
15. Nelson TW, Rose SA (2016) Controlling hard zone formation in friction stir processed HSLA steel. J Mater Process Technol 231:66–74
16. Taysom BS, Sorensen CDCD, Taysom BS, Sorensen CDCD (2019) Advances in signal processing for friction stir welding temperature control. Friction Stir Weld Process X:135–147.

Part IV
Dissimilar

Dissimilar Joining of ZEK100 and AA6022 for Automotive Application

Hrishikesh Das, Piyush Upadhyay, Shank S. Kulkarni, and Woongjo Choi

Abstract Mg-Al joining is problematic due to dissimilar high-temperature flow characteristics that often result in defects and brittle intermetallic compounds (IMC) at the interface. Process development on lap joining of 1.27 mm 6022 (top sheet) and 1.5 mm ZEK100 (bottom sheet) intended for an automotive door application is presented. Process development was conducted in linear welding using power control (modulating torque and spindle speed) with multiple tool designs and welding parameters. Process variables were correlated to micro- and macro-structure and mechanical properties. A 2D computational model was created to understand the failure mechanism and contribution of mechanical interlocking towards joint strength. Joint efficiency of 48% (compared to base 6022 Al) at a welding speed of 0.625 m/min has been demonstrated thus far. For a thin (<2 mm) automotive sheet with a softer material as a bottom sheet, adequate control of process variables is critical to maintain weld stability and avoid excessive IMC layer at the interface. A combination of mechanical anchoring and metallurgical reaction is likely responsible for bonding.

Keywords Dissimilar joining · FSW · Cohesive zone model

Introduction

With growing demand for lightweight vehicles in recent times, aluminum (Al) and magnesium (Mg) alloys are being considered for a variety of applications in automotive industries. This has precipitated a need for an effective, low-cost method to join Mg and Al alloys that meet the multi-material design needs. Several challenges exist towards this goal. Conventional fusion-based joining approaches often result in voids and thick and brittle intermetallic compounds. Friction stir welding,

H. Das · P. Upadhyay (✉) · W. Choi
Applied Materials and Manufacturing Group, Pacific Northwest National Laboratory, Richland, USA
e-mail: Piyush.Upadhyay@pnnl.gov

S. S. Kulkarni
Advanced Computing, Mathematics and Data Division, Pacific Northwest National Laboratory, Richland, USA

Y. Hovanski et al. (eds.), *Friction Stir Welding and Processing XI*, The Minerals, Metals & Materials Series, https://doi.org/10.1007/978-3-030-65265-4_11

a solid-state shear-assisted joining process, has a potential to minimize these problems. Joining of Mg and Al pairs has been subject to multiple study. Most of the work reported have used butt- and spot-welding configurations [1–8]. Friction stir lap welding of Al/Mg has also been explored recently but progress is limited. Ji et al. [9] reported lap welds between 3 mm 6061 and 3 mm AZ31 using stationary shoulder. They reported that (a) with increasing the weld length tunnel defects approaching due to the poor flow ability of Mg, (b) lap shear strength is significantly higher when Mg is placed on the top, (c) Al_3Mg_2 and $Al_{12}Mg_{17}$ type of intermetallics (IMCs) form for Mg on top and Al on top lap joint cases, respectively. Shen et al. [4] also found better mechanical performance for Mg (top sheet)/Al joint compared to Al (top sheet)/Mg. Mohammadi et al. [2] and Chen et al. [3] noticed that welding speed has high impact to determine the joint strength for Al/Mg dissimilar joint. Kostka et al. [1] concluded that load bearing capacity was determined by the 1 μm IMC layer consisting of $Al_{12}Mg_{17}$ phase. In a separate study, Mohammadi et al. [10] noted that tapered threaded pin is beneficial to achieve higher joint strength. Liang et al. [11] detected several weak links in Al/Mg FSW joints, namely, (a) bottom interface, (b) banded zone, and (c) softened Al in the retreating side.

While several groups have shown the preliminary potential for FSW of Al/Mg dissimilar joints with dissimilar thicknesses, most of the work have been done on 2 mm or thicker sheets. Tool geometry and weld schedule to attain a viable, defect-free weld in thin sheet Mg to obtain continuous repetitive linear welds with reasonable joint strength is lacking.

The objective of the present work is to fill this gap by developing welding parameters, tool geometry, and control strategies for high-speed welding speed (> 0.5 m/min) with potential of commercialization. SEM–EDS-based microstructural characterization was exhibited at the vicinity of the interface. 2D computational model was created to understand the failure mechanism and contribution of mechanical interlocking towards joint strength.

Experimental Methods

Sheets of 6022 AA (1.27 mm thick), ZEK 100 (1.5 mm thick) were used for lap joining. All the lap joints were performed in spindle power control mode developed at Pacific Northwest National Laboratory (PNNL). The power transferred into the weld zone was evaluated by measuring the input mechanical power. The mechanical input power is calculated as follows:

$$P = \tau \times \omega \times \frac{2\pi}{60}$$

where τ and ω are the spindle torque, expressed in units of N-m, and the spindle speed, expressed in units of RPM, respectively. In order to transfer a constant power

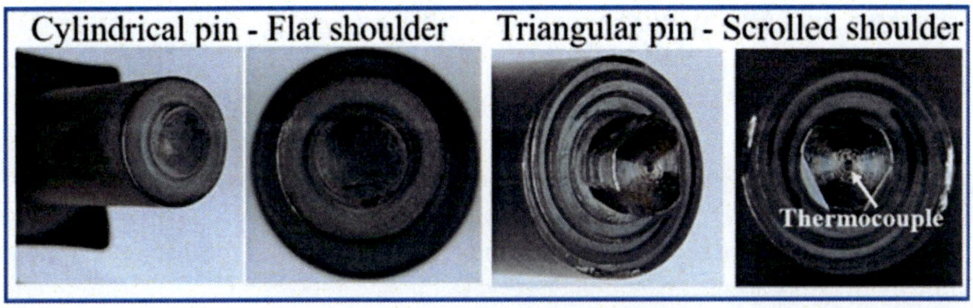

Fig. 1 Cylindrical pin with flat shoulder and triangular pin with scrolled shoulder. (Color figure online)

input throughout the linear welds, a closed-loop algorithm was implemented. The mechanical power input was controlled by modulating torque from the spindle motor and monitoring spindle speed of the motor using an encoder sensor. The torque modulating method shows less than 1% of error on power control throughout the linear weld.

Cylindrical pin with flat shoulder and triangular pin with scrolled shoulder were used for this study as shown in Fig. 1. To measure the temperatures at the weld interface, a thermocouple was inserted through the tool body and appeared through the pin tip as shown in Fig. 1. The microstructures at the weld interface were characterized using SEM with EDS. In accord with Standard ASTM D1002-10, transverse lap shear tensile samples 30 mm wide × 150 mm long were cut by electrical discharge machining. Unguided lap shear tensile tests were conducted with an initial strain rate of 0.001 s^{-1} using an Instron 8860 universal tensile testing machine.

A two-dimensional finite element model (FEM) was implemented in ABAQUS [12] to study the contribution of hook feature on overall weld strength in lap shear loading. Figure 2 shows the overall geometry and boundary conditions used for FE model. The cross section is taken in transverse direction of weld with assumption of plane strain. Left end of Al sheet is fixed while uniform displacement is applied on the right end of Mg sheet as shown in Fig. 2. A characteristic hook feature obtained from a polished sample is introduced along the interface of weld. Simulation is repeated without hook feature and its impact on strength of joint is analyzed.

Fig. 2 Schematic of finite element model geometry along with boundary conditions for lap shear test of joint. (Color figure online)

Results and Discussion

Dissimilar thickness of 6022 Al and ZEK Mg along with Mg sheet on the bottom in lap configuration made the process development more complicated. Conventional route of process development using rpm control was not successful. Lack of material consolidation and defects and surface issues were observed especially when weld ran longer than 150 mm. Interface of the joints exhibited cavities for different combination of process parameters with RPM control mode as shown in Fig. 3. Subsequently, power control mode was used with two different tool geometries (shown in Fig. 1a, b). Process parameters corresponding to different tool geometries are listed in Table. 1. Figure 4a, b shows that repetitive long welds were executed without surface defects for both the tool geometries. It is interesting to note that different tool geometries reacted differently with a specific spindle power and as a result commanded power to obtain similar crown surface vary. Nevertheless, peak measured pin tip temperature (~325–340 °C) is almost similar for both the cases irrespective of difference in power profile. Whereas Z force is greater with triangular pin (~10 kN) compared to cylindrical (~8 kN) which is probably due to the aggressive scrolled shoulder in the triangular tool (Fig. 5a, b). Interface for both the joints is shown in Fig. 6a, b, respectively. Figure 6a indicates that some defects still exist with the cylindrical tool. Whereas no macro-defects persist with triangular tool. Two hook features (inset of Fig. 6b: b1 and b2) on the either side of the weld nugget are observed at the interface with triangular tool (Fig. 6b) while a single prominent hook feature is produced with cylindrical tool. Lap shear tensile results and corresponding DIC for both the joints are shown in Fig. 7a, b, c, respectively. For cylindrical tool, the crack path directly

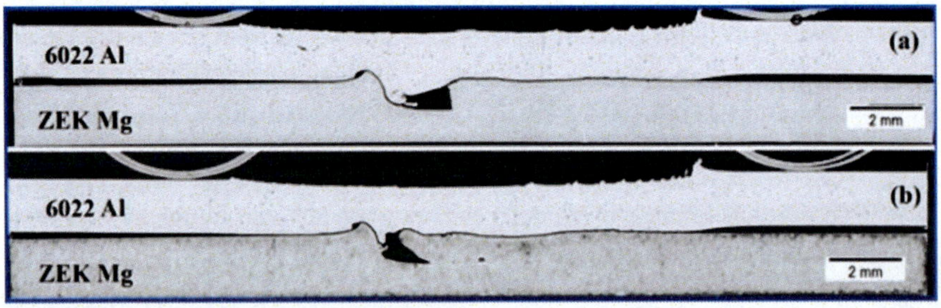

Fig. 3 Interface of the Al/Mg joints with RPM control mode; **a** 1950 RPM and 1 m/min, **b** 1000 RPM and 0.5 m/min. (Color figure online)

Table 1 Process parameters

Tool geometries	Power (kW)	Travel speed (m/min)	Depth of penetration (mm)	Tilt angle (°)
Cylindrical	2.2	0.625	2.6	1
Triangular	1.9	0.625	2.3	1

Fig. 4 Weld surface for **a** cylindrical tool and **b** triangular tool. (Color figure online)

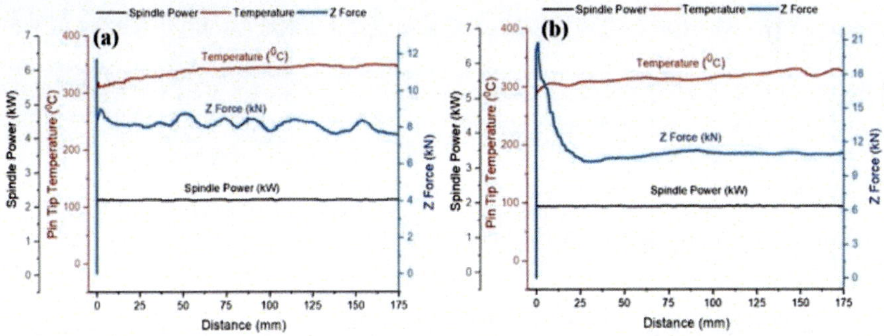

Fig. 5 Spindle power, pin tip temperature, and Z force for **a** cylindrical tool and **b** triangular tool. (Color figure online)

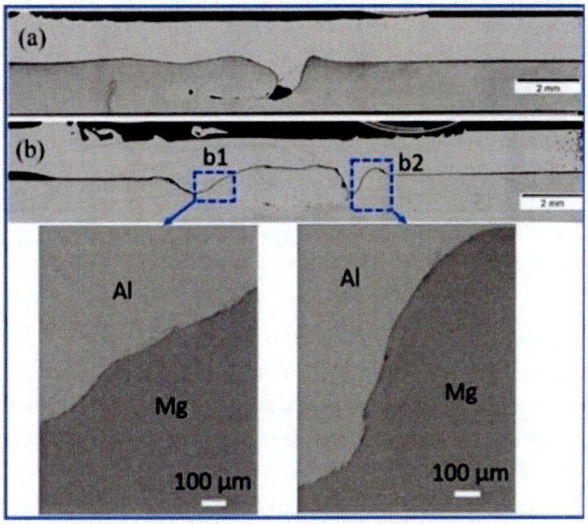

Fig. 6 Interface of Al/Mg joint with **a** cylindrical and **b** triangular tool. (Color figure online)

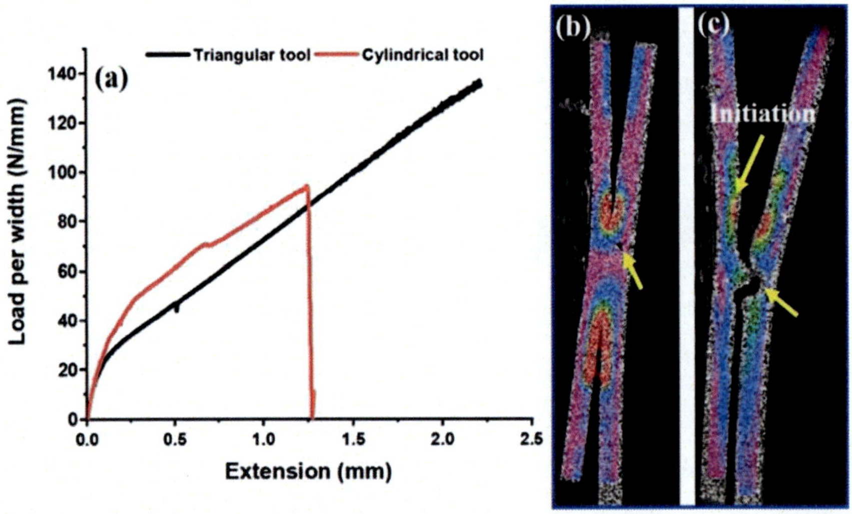

Fig. 7 **a** Lap shear tensile comparison, crack initiation and propagation through DIC for **b** cylindrical and **c** triangular tool. (Color figure online)

moves through the cavity at the interface as marked by yellow arrow in Fig. 7b. However, for triangular tool, crack first initiates from the small hook and then crack finally propagates through the large hook. To represent the exact hook geometry in computational model, cross-section image was used which is shown in Fig. 6. The coordinates of hook boundary were traced. A smooth spline was fitted along those coordinates using Python scripts in ABAQUS model as shown in Fig. 8a, c. The finite element mesh was refined near the hook feature to increase the accuracy of simulations as shown in Fig. 8b, d. Cohesive Zone Model (CZM) was used in this study to define the interface bonding between Al and Mg sheets after welding [13]. Continuum response of both sheets is modeled by elastic–plastic material model [14]. The elastic and plastic material parameters are taken from literature [15].

To introduce damage in the continuum material, a Johnson–Cook damage model [16] was used for both Al and Mg sheets. The J-C fracture model states that the fracture strain depends on the stress triaxiality ratio. Once the damage initiation criterion is reached in any given finite element, its loading capacity would be decreased until a failure criterion (fracture energy for this work) is reached. The peak load obtained in both cases of lap shear test (with and without hook feature) was compared. Figures 9 and 10 show stress distribution over the weld area, and ductile damage during the peak load for weld with cylindrical tool and triangular tool, respectively. The contribution of hook feature in overall strength of weld is found to be 24% in case of weld with cylindrical tool and 18% in case of triangular tool.

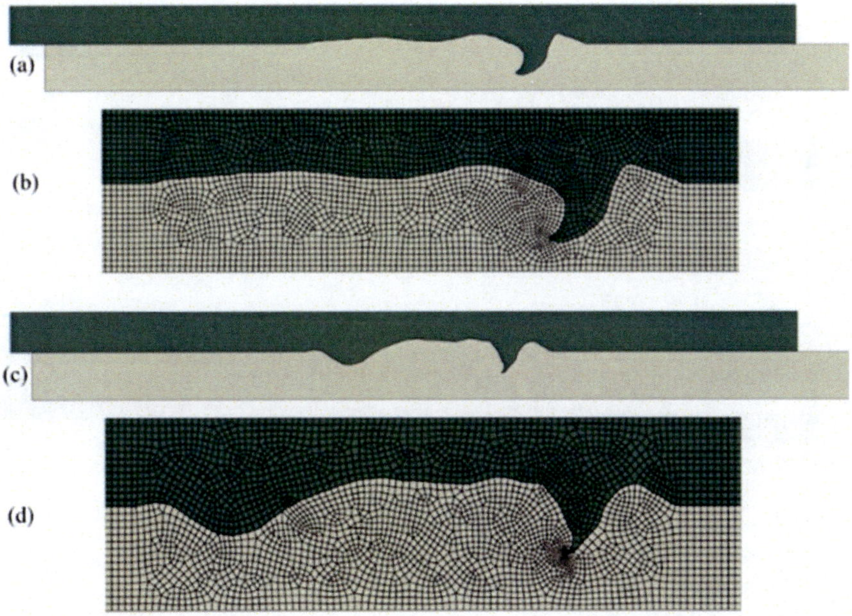

Fig. 8 **a** Representation of the hook feature geometry in computational model, **b** finite element mesh near the hook feature for weld with cylindrical tool, **c** representation of the hook feature geometry in computational model, and **d** finite element mesh near the hook feature for weld with triangular tool. (Color figure online)

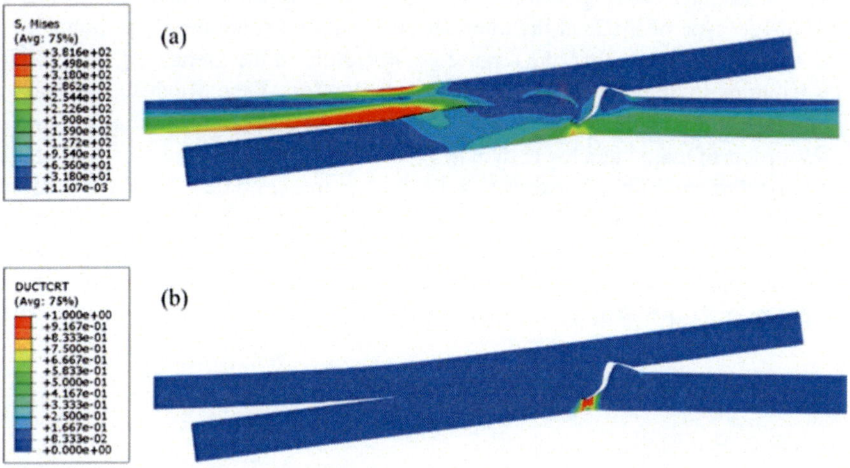

Fig. 9 **a** Stress distribution over the weld area and **b** ductile damage during the peak load for weld with cylindrical tool. (Color figure online)

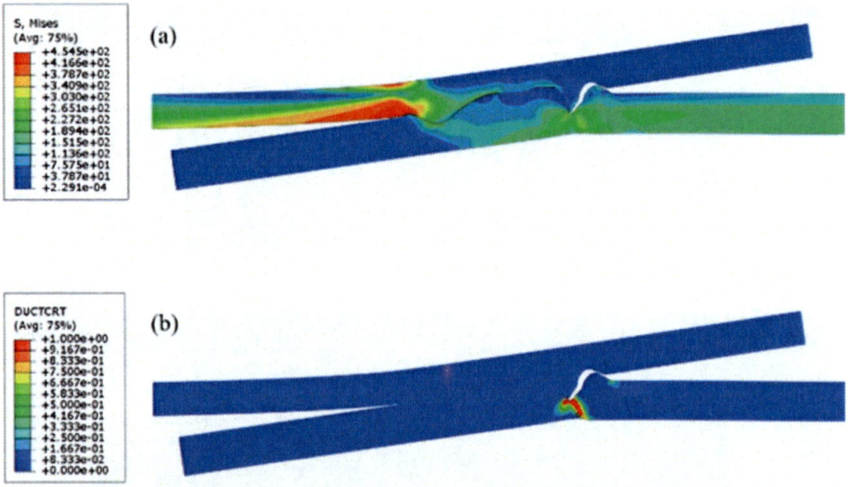

Fig. 10 **a** Stress distribution over the weld area and **b** ductile damage during the peak load for weld with triangular tool. (Color figure online)

Interface Characterization

SEM with EDS line scan shows that the interface layer thickness is around 500–600 nm (Fig. 11a). Al and Mg peaks also match indicating formation of an Al/Mg IMC at the interface (Fig. 11b). Point analysis suggests the formation of Al_3Mg_2 and $Al_{12}Mg_{17}$ type of IMCs at the interface. It should be noted here that the measured temperature is lower than the reported temperature in the literature [2–5]. While it is tempting to draw from the existing knowledge for Al and Mg diffusion across an interface at high temperature and pressure, the time scale of the joining process is a few orders of magnitude lower than in a typical diffusion experiment. For example, a 1 mm interface is directly exposed to the heat source for 0.02 s. Also, the total sheet thickness is 2.77 mm, thus much higher cooling rate would be expected. Therefore, the interfacial chemistry is a more localized phenomenon and could vary slightly with location as well. A high welding speed and short diffusion time explain a nano IMC layer (500–600 nm) at the interface.

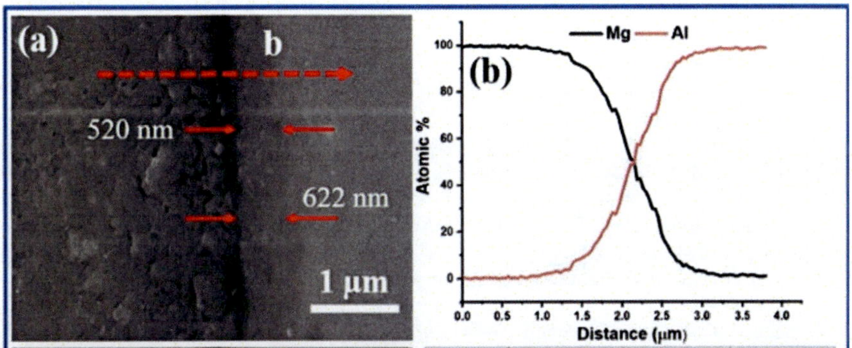

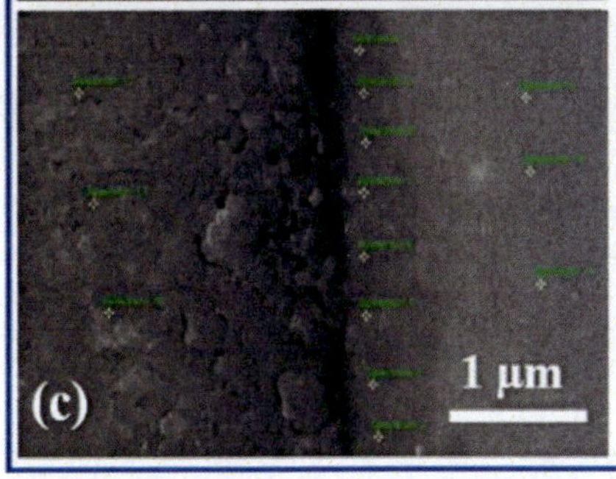

Spectrum	Mg (atom. %)	Al (atom. %)	Intermetallic
1	40.8	59.1	Al_3Mg_2
2	56.3	43.6	$Al_{12}Mg_{17}$
3	40.7	59.2	Al_3Mg_2
4	58.4	41.5	$Al_{12}Mg_{17}$
5	57.2	42.7	$Al_{12}Mg_{17}$
6	53.4	46.5	$Al_{12}Mg_{17}$
7	46	53.9	Al_3Mg_2
8	40.5	59.4	Al_3Mg_2
9	1.7	98.2	Al Side
10	1.5	98.4	Al Side
11	0.90	99.1	Al Side
12	99.5	0.40	Mg Side
13	99.7	0.23	Mg Side
14	98.8	1.1	Mg Side

Fig. 11 **a** Interface with location of EDS line scan, **b** EDS line scan analysis plot, **c** location of EDS point analysis and respective vol. fraction with probable IMCs. (Color figure online)

Conclusion

Power control mode has been successfully implemented for lap joining of dissimilar Al/Mg. Both mechanical (hooking) and metallurgical bonding contributed towards the intimate contact at the interface. Per preliminary modeling work the contribution of hook feature in overall strength of weld is found to be somewhere between 18 and 24% and remaining joint strength was derived from the nanoscale (500 ± 100 nm) thin layer of Al/Mg IMC. Computational model for failure mode and stress distribution shows good agreement with the experimental observation.

Acknowledgements PNNL is operated by Battelle Memorial Institute for the U.S. Department of Energy under contract DE-AC05-76RL01830. This work was sponsored by the DOE- EERE, Vehicle Technology under project titled "Phase Field Modeling of Corrosion for Design of Next-Generation Magnesium-Aluminum Vehicle Joints" partnership of Worcester Polytechnic Institute, Oak Ridge National Lab and Magna. We are thankful to Daniel Graff to assist during welding; Anthony Guzman for metallographic preparation; Tim Roosendaal, Ethan Nickerson, and Robert Seffens for mechanical testing and DIC analysis; and Joshua Silverstein for SEM characterization. The authors appreciate material support provided by Tim Skszek, Magna.

References

1. Kostka A, Coelho RS, Santos J dos, Pyzall AR (2009) Microstructure of friction stir welding of aluminum alloy to magnesium alloy. Scr Mater 60 : 953–956. https://www.sciencedirect.com/science/article/pii/S1359646209001080
2. Mohammadi J, Behnamian Y, Mostafaei A, Gerlich AP (2015) Tool geometry, rotation and travel speeds effects on the properties of dissimilar magnesium/aluminum friction stir welded lap joints. Mater Des 75: 95–112. https://www.sciencedirect.com/science/article/pii/S0261306915001004
3. Chen YC, Nakata K (2008) Friction stir lap joining aluminum and magnesium alloys. Scr Mater 58: 433–436. https://www.sciencedirect.com/science/article/pii/S1359646207007713
4. Shen JJ, Li Y, Zhang T, Peng D, Wang D, Xu N (2015) Preheating friction stir spot welding of Mg/Al alloys in various lap configurations. Sci Technol Weld Join 20 (1): 1–10. https://doi.org/10.1179/1362171814Y.0000000248
5. Kwon YJ, Shigematsu I, Saito N (2008) Dissimilar friction stir welding between magnesium and aluminum alloys. Mater Lett 62: 3827–3829. https://www.sciencedirect.com/science/article/pii/S0167577X08004369
6. Buffa G, Baffari D, Di Caro A, Fratini L (2015) Friction stir welding of dissimilar aluminum-magnesium joints: sheet mutual position effects. Sci Technol Weld Join 20 (4): 271–279. https://doi.org/10.1179/1362171815Y.0000000016
7. Venkateswaran P, Reynolds AP (2012) Factors affecting the properties of Friction Stir Welds between aluminum and magnesium alloys. Mater Sci Eng A 545: 26–37. https://www.sciencedirect.com/science/article/pii/S0921509312002870
8. Fu BL, Qin GL, Li F, Meng XM, Zhang JZ, Wu C S (2015) Friction stir welding process of dissimilar metals of 6061-T6aluminum alloy to AZ31 B magnesium alloy. J Mater Process Technol 218: 38–47. https://www.sciencedirect.com/science/article/pii/S0924013614004634
9. Ji S, Li Z, Zhang L, Zhou Z, Chai P (2016) Effect of lap configuration on magnesium to aluminum friction stir lap welding assisted by external stationary shoulder. Mater Des 103: 160–170. https://www.sciencedirect.com/science/article/pii/S0264127516305408
10. Mohammadi J, Behnamian Y, Mostafaei A, Izadi H, Saeidd T, Kokabi AH, Gerlich AP (2015) Friction stir welding joint of dissimilar materials between AZJ 1 B magnesium and 6061 aluminum alloys: microstructure studies and mechanical characterizations. Mater Charact 101: 189–207. https://www.sciencedirect.com/science/article/pii/S1044580315000108?via%3Dihub
11. Liang ZY, Chen K, Wang XN, Yao JS, Yang Q, Zhang LT, Shan AO (2013) Effect of tool offset and tool rotational speed on enhancing mechanical property of Al/Mg dissimilar FSW joints. Metal Mater Trans A 44A: 3721–3731. https://doi.org/10.1007/s11661-013-1700-4
12. Smith M (2009) ABAQUS/Standard User's Manual, Version 6.9, Dassault Systems Simulia Corp, United States
13. Elices MG, Guinea GV, Gomez J, Planas J (2002) The cohesive zone model: advantages, limitations and challenges. Eng Fracture Mech 69(2):137–63. https://www.sciencedirect.com/science/article/pii/S0013794401000832
14. Khan AS, Huang S (1995) Continuum theory of plasticity. Wiley, New York. https://www.wiley.com/en-us/Continuum+Theory+of+Plasticity-p-9780471310433
15. Wang T, Tamayo DR, Jiang X, Kitsopoulos P, Kuang W, Gupta V, Barker E, Upadhyay P (2020) Effect of interfacial characteristics on magnesium to steel joint obtained using FAST. Mater. Des. 28:108697. https://www.sciencedirect.com/science/article/pii/S0264127520302318
16. Johnson GR, Cook WH (1998) Fracture characteristics of three metals subjected to various strains, strain rates, temperatures and pressures. Eng Fract Mech 21(1):31–48. https://www.sciencedirect.com/science/article/pii/0013794485900529

Fracture Mechanics Approach to Improve Fatigue Strength of a Dissimilar Metal T-Lap Joint by Friction Stir Welding

Masakazu Okazaki, Hao Dinh Duong, and Satoshi Hirano

Abstract Friction Stir Welding (FSW) is expected as one of the industrial solutions to fabricate high-performance dissimilar metal welds consisting of materials to which conventional fusion welding process(es) is (are) difficult to apply. However, the strength of FSWed T-lap joint, especially for the fatigue failures, has not been satisfactory so far, due to the formation of some undesirable defects in the joint, e.g., tunnel, kissing bond, zigzag line, bonding line, oxide line defects, etc., those may directly degrade the mechanical properties of the joint. In this work, a trial was made to minimize the defect size in the dissimilar metal FSWed T-lap joint between AA7075 and AA5083, by employing a double stir welding method. The fatigue strength of the joint was also explored, paying attention to the fatigue crack nucleation sites and the morphologies. Some fracture mechanics strategies are proposed to improve the fatigue strength associating with enough rupture ductility.

Keywords Dissimilar FSW T-lap joint · Double-pass FSW · Bonding interface · Fatigue strength · Kissing bond defects · Crack initiation · Digital Image Correlation (DIC) technique

Introduction

Fusion welding technology has been widely applied for various materials. Some methods are known as arc welding, gas welding, radiant energy welding, resistance welding, etc. In these methods, the materials are heated up to a melting temperature

M. Okazaki (✉)
Nagaoka University of Technology, 1603-1 Kamitomioka-machi, Nagaoka-shi, Niigata 940-2188, Japan
e-mail: okazaki@mech.nagaokaut.ac.jp

H. D. Duong
Nagaoka University of Technology, 1603-1 Kamitomioka-machi, Nagaoka-shi, Niigata 940-2188, Japan

S. Hirano
Hitachi Research Lab, Hitachi 319-1292, Japan

Y. Hovanski et al. (eds.), *Friction Stir Welding and Processing XI*, The Minerals, Metals & Materials Series, https://doi.org/10.1007/978-3-030-65265-4_12

before they diffuse each other [1–5]. There are some cases where these are generally difficult to be applied. For example, the precipitation hardening of alloys produced by conventional welding method frequently causes hot cracking, distortion, and high residual stress [1–5]. Friction Stir Welding (FSW) that was invented by The Welding Institute (TWI) in 1991 can supply one of the solutions [1]. Since this technique uses frictional heat induced by rotating tool to weld materials, it enables us that the welding temperature during FSW process is lower than the melting temperature of base materials. The FSW is expected to be applied to produce a large amount of various industries such as automotive, shipbuilding, high-speed train manufacturing, aerospace, and railway, where the T-joints composed of stringer and skin plates are the common junctions. The FSW must be also useful to produce the joints between dissimilar metals with different physical and mechanical properties.

The main purpose of this work is to investigate the interface morphology and fracture behavior of the FSWed dissimilar metal T-lap joint between AA7075 and AA5083 aluminum alloys, those make platform for the improvement of the mechanical properties of the joints. Special attention was paid to the FSW conditions to improve the strength associating with enough rupture ductility, and to explore some academic backgrounds on how and why the joint strength was affected by many FSW process parameters.

Defects in the FSWed T-Joint

The FSWed T-Lap joint was fabricated according to a process illustrated in Fig. 1, where the pin was rotating from the top of skin plate. The tilt angle of the pin was set up by 2.0 degree in all cases. Many FSW variables were tried in this work; they are rotational speed, tool offset, welding speed, and pin depth. Those conditions will be given on such occasion to present each experimental result hereinafter. More or less, some types of defects were often found in the joint, as classified in Fig. 1c, where the terms "RS" and "AS" denote the retreating and advancing sides in FSW, respectively. Here, the interface located outside stir zone is referred to as kissing

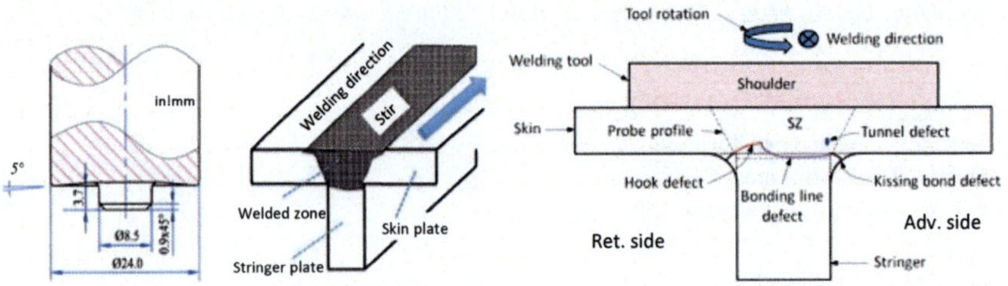

(a) Pin geometery (b) T-lap joint by FSW (c) Classification of typical defects

Fig. 1 Fabrication of T-lap joint by FSW. (Color figure online)

bond defects [1–3]. The interface is located inside the SZ where hook geometry is referred to as a hook defects that are often observed in the FSWed lap joint. The other interface located inside the SZ with oxide film is referred as bonding line defects.

Mechanical Properties of T-Lap Joints

A single-pass FSW was used to produce the T-lap joint consisting of two aluminum alloys, for the stringer Alloy 7075-T651 (denoted as AA7075) and for the skin plate 5083-H116 (denoted as 5083) those had the dimensions of 300 × 50 × 8.2 mm and 300 × 150 × 3.0 mm, respectively. The welding speed was arranged from 50 to 200 mm/min and the rotational speed was kept constant at 400 rpm.

The performance of the joint was evaluated by two types of mechanical tests, extracting small samples from the T-joint: one is the tensile test along the skin part (denoted as "skin test") and the second is along the stringer part (denoted as "stringer test", see Fig. 2).

The effect of welding speed on the formation of four types of defects in T-lap joints is quantitatively summarized in Figs. 3 and 1b. Here, the defect size was directly measured based on oxide film on the cross section of the specimens by means of the optical microscope. It is found from Fig. 3 that the welding speed played a significant role in the defect size. The tunnel and bonding line defects were drastically changed by controlling welding speed. Increase in welding speed was useful to reduce hook defects size.

Meanwhile, the KB defect which formed under all welding conditions was insensitive to welding speed. It seems that the specimen produced at the welding speed of 100 mm/min showed the best interface with small defect size. Even in this case the

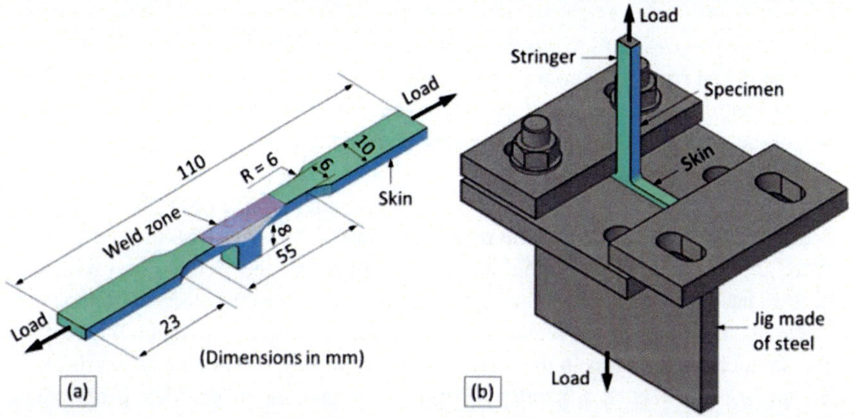

Fig. 2 Two kinds of tests to evaluate the FSW properties. **a** skin test **b** stringer test. (Color figure online)

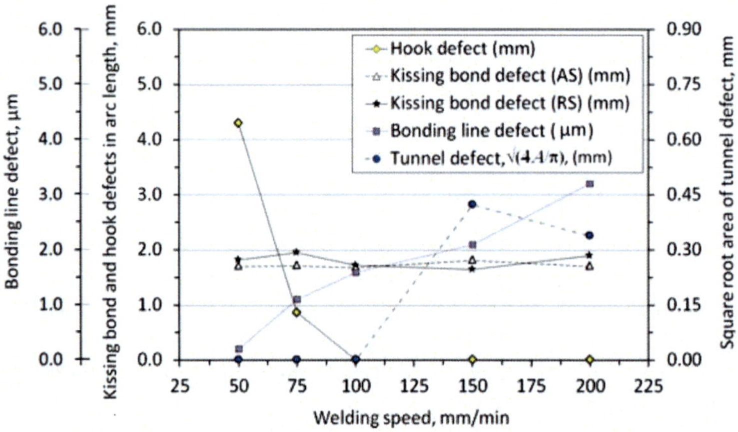

Fig. 3 Influence of welding speed on the formation of each type of defect. (Color figure online)

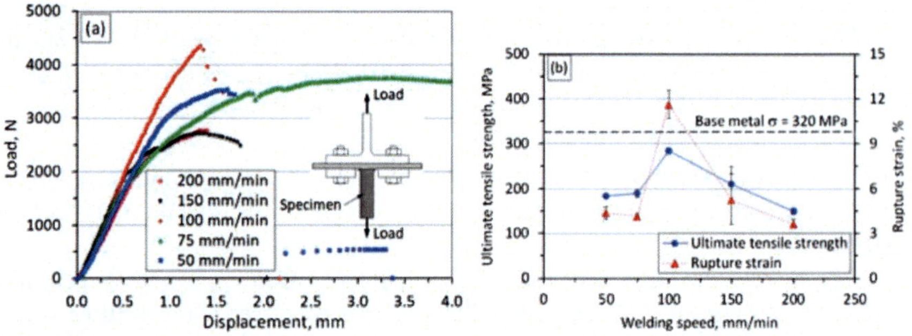

Fig. 4 Stringer test results. **a** load–displacement curves and **b** maximum load and displacement at maximum load. (Color figure online)

joint suffered from low rupture strain in the stringer test, as shown in Fig. 4. Thus, it seems that the applicability of the traditional single-pass FSW is definite to achieve well-balanced strength between the skin tests and the stringer test and to control the weld interface morphologies.

Taking account of above limitations by the single-pass FSW, both the double-pass FSW which can induce the reversed material flows and the tool offset method were applied, see Fig. 5. Note that there is tool offset between the first pass and the second pass, and the second pass has a role to reduce the defect size which have been nucleated by the first pass welding process. The application of double FSW changed the defect morphologies in the welds, as shown in Fig. 6. The tool offset towards the advancing side was significantly helpful to improve the mechanical properties than that towards the retreating side. This mechanical background will be discussed in the next section.

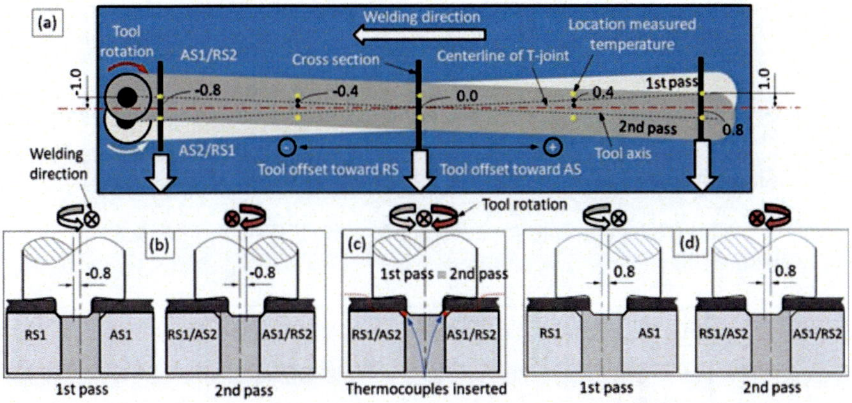

Fig. 5 Illustration of double FSW associating with tool offset. (Color figure online)

The effectiveness of the double-pass FSW was also confirmed by the digital image correlation technique, indicating that the plastic deformation was inhomogeneous associating with significant strain localization around the KBs in most cases. The noteworthy was that the strain localization was relaxed in the specimens by offsetting the tool probe toward the AS with 0.8 mm, where the KB defect size was also minimized.

Following these successful results, the fatigue properties of the double-pass FSW joint were evaluated, see Fig. 7, which indicates the fatigue strength under the skin loading reached relatively 80% to that of the two base metals. However, the strength under stringer loading mode was still lower.

Fracture Mechanics Indication to the FSW Process Variables

It is meaningful to discuss the mechanical interpretations why the strength of FSW joint was significantly influenced by many welding variables presented in the previous sections, together with what are roles of them in changing the mechanical properties. The state of the joint in the skin and stringer tests to which external load is applied can be modeled by Figs. 8 and 9, respectively. On the other hand, it has been well known that the Stress Intensity Factor (SIF), a fracture mechanics parameter which is given in terms of defect size, external remote stress, and crack geometry, is very useful parameter to discuss fractures. Fracture mechanics teaches that the magnitude of SIF of the single-edge crack is higher than those of the double-edge and the dispersed crack. Here it is worthy to remind the double-pass welding dispersed the defects into the AS and RT sides comparably, compared with the single pass (Fig. 9b, c).

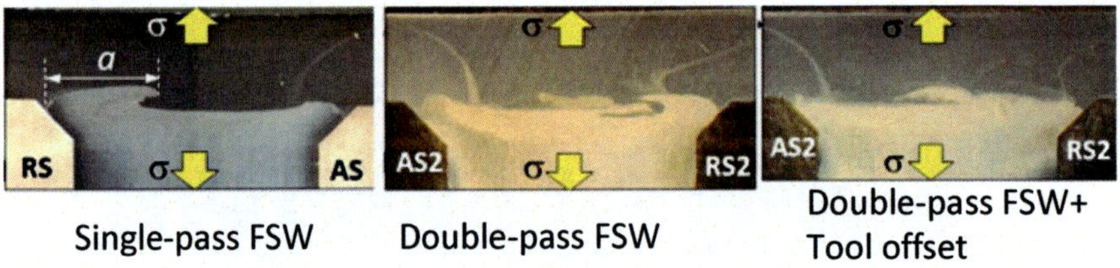

(a) Change in interface morphology by FSW conditions.

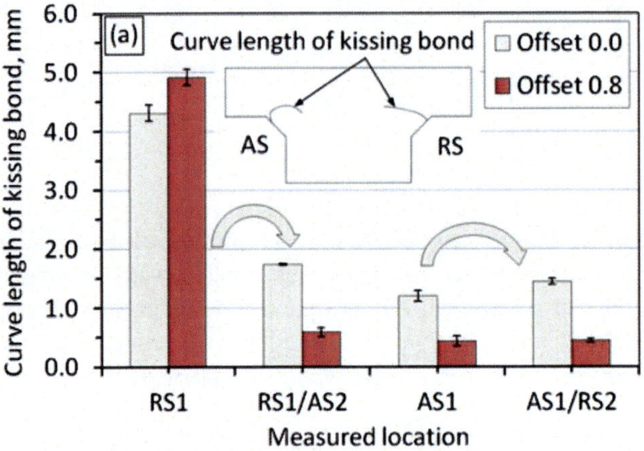

(b) Change in interface defect morphology

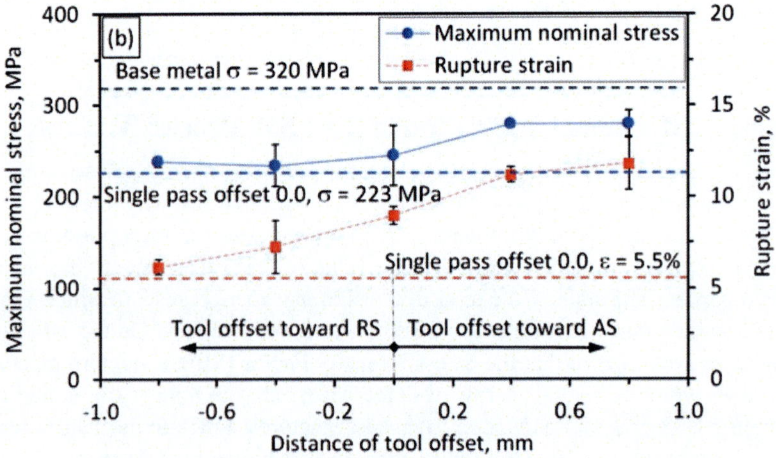

(c) Change of strength in the skin test.

Fig. 6 Effectiveness of the double FSWs associating with tool offset. (Color figure online)

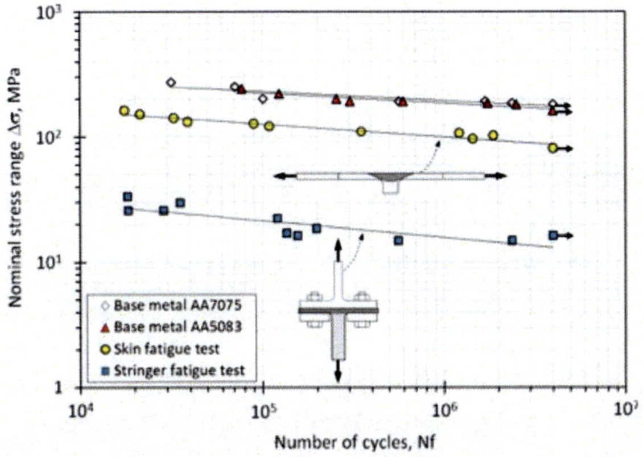

Fig. 7 Fatigue strengths of the double FSWs under skin and stringer loadings. (Color figure online)

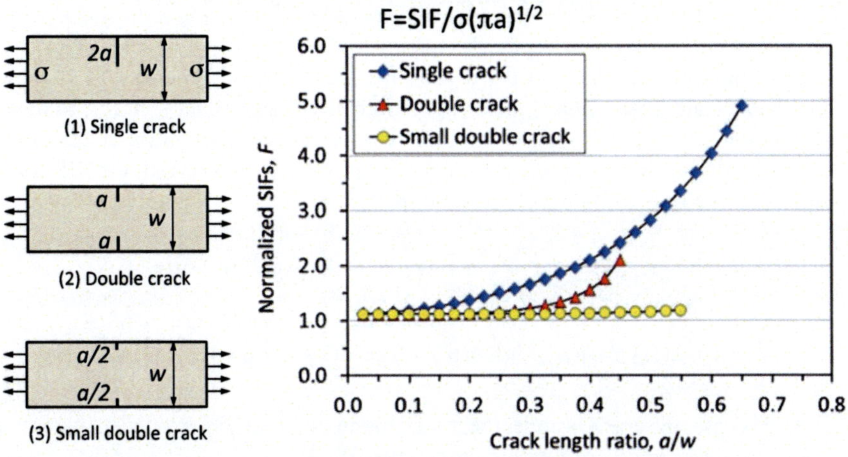

Fig. 8 SIF strongly affected by crack configurations. (Color figure online)

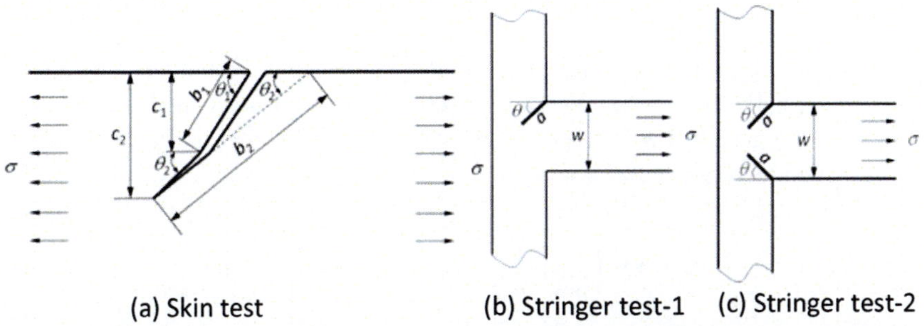

Fig. 9 Simplified model in the skin and stringer tests for the defected joint

The defect geometry nucleated in the present work is not so simple as in the above simplification, but it is often inclined, or kinked, to loading axis (Fig. 9). Especially the crack seen in the single FSW T-joint was significantly kinked and curved at the RS side (Fig. 9a). The SIFs of the crack in Fig. 9a can be approximated for mode I and mode II (denoted by K_I and K_{II}) as follows [5, 6]:

$$K_I = F \times \left[(\sin\theta_2)^2 \times \sigma\right] \times (1.12)^2 \times \frac{\sqrt{\pi b_2}}{\sqrt{\sin\theta_2}} \tag{1}$$

$$K_{II} = F \times (\sin\theta_2 \cos\theta_2 \times \sigma) \times \frac{1}{1.12^2} \times \frac{\sqrt{\pi b_2}}{\sqrt[4]{\sin 2\theta_2}} \tag{2}$$

where the factor F is determined as a function of $\lambda = b_2/w$, which is given by.

$$F = \frac{1}{\sqrt{\pi}}\left(1.99 - 0.41\lambda + 18.7\lambda^2 - 38.5\lambda^3 + 54\lambda^4\right) \tag{3}$$

It is important to note that the K_{I} and K_{II} are sensitively changed by the crack length "b_2" and little by "b_1", when the length c_2 is longer than $\gtrsim 1.2c_1$, following the reference [6]. This suggests the FSW condition may change the magnitude of SIF through the change in crack configurations. On the other hand, the SIF for the kinked edge crack in Fig. 9b is given by [6, 7]

$$K_I = F \times Y_I \sigma \sqrt{\pi a} \quad Y_I = 1.12^2 (\sin\theta)^2 / \sqrt{\sin\theta} \tag{4}$$

$$Y_I = 1.12^2 (\sin\theta)^2 / \sqrt{\sin\theta} \quad Y_{II} = \frac{1}{1.12^2} \sin\theta \cos\theta / \sqrt[4]{\sin 2\theta} \tag{5}$$

In this case, the crack is under the mixed mode, for which an equivalent parameter, Y_{eq}, is effective upon considering the fracture [8, 9].

$$Y_{eq} = \sqrt{(Y_I)^2 + (Y_{II})^2} \tag{6}$$

In this work, the parameters $1/(F \times Y_I \sqrt{a})$ and $1/(F \times Y_{II} \sqrt{a})$ are called by "Geometrical resistance factor of defects (GRFD)". On the other hand, for the double-edged crack which was formed under the double FSW process (Fig. 9c), the ***F*** in Eqs. (4) and (5) can be replaced by.

$$F = \left[1 + 0.122\left(\cos\frac{\pi\xi}{2}\right)^4\right] \times \sqrt{\frac{2}{\pi\xi}\tan\frac{\pi\xi}{2}} \quad \text{with } \xi = 2a/w \tag{7}$$

For the SIF of the single-edged crack (Fig. 8b) the F is given by.

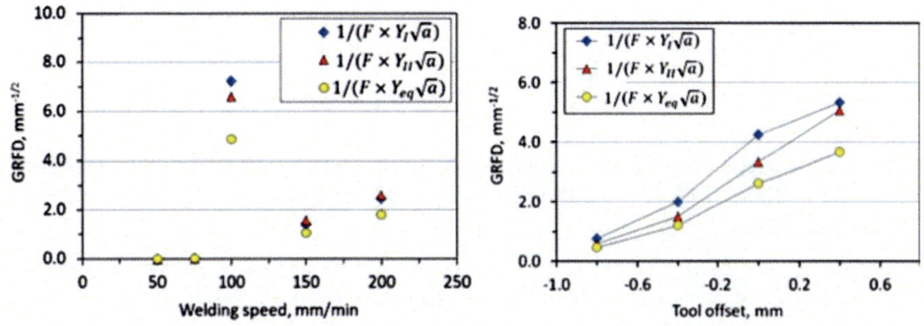

(a) GRFD affected by welding speed. (Single-FSW, Skin test)

(b) GRFD affected by tool offset (double-FSW, skin test)

Fig. 10 Change in GRFD by changing FSW conditions. (Color figure online)

$$F = \frac{1}{\sqrt{\pi}}\left(1.99 - 0.41\lambda + 18.7\lambda^2 - 38.5\lambda^3 + 54\lambda^4\right) \quad \text{with } \lambda = a/w \tag{8}$$

The larger value of GRFD corresponds to higher strength. The effects of welding speed and probe length on the GRFD in the skin test are shown in Fig. 10, respectively. Note that the GRFD showed the peak value at the welding speed of 100 mm/min, explaining well the experimental results that the strength showed maximum under this condition. The similar good agreement is also found in Fig. 10b. Figure 10 explains well the difference in the strength between the single and double FSW joints. It is also important to note that the change in FSW conditions changes the interface geometry resulting in the change in inclined angles, $\boldsymbol{\theta}, \boldsymbol{\theta}_1$ and $\boldsymbol{\theta}_2$ appearing in Eqs. (1) through (5).

Conclusion

The effects of FSW parameters, welding speed, tool offset, pin depth, and application of double FSW on the mechanical properties were explored by the skin test, stringer test, and the fatigue tests. It was found that the performance of dissimilar T-joint might be surely optimized by controlling these factors. There was an optimum welding speed in skin test of the single FSW. The application of double FSW changed the defect morphologies in the welds, e.g., while the defect was predominant at the RS side in the single FSW, they are comparably dispersed at the RS and AS sides in the double FSW. In the latter, the defect size was also reduced. Based on fracture mechanics a new concept has been proposed to make the analysis of the strength of dissimilar FSWed T-lap joints, by introducing the "GRFD" parameter. This concept could provide reasonable explanations to the effect of FSW conditions.

Acknowledgements A part of this work is based on results obtained from a future pioneering project commissioned by the New Energy and Industrial Technology Development Organization (NEDO).

References

1. Thomas WM, Nicholas ED, Needham JC, Murch MG, Temple-Smith P, Dawes CJ (1991) Friction-stir butt welding. In: International patent application PCT/GB92/02203 and GB patent application 9125978.8, UK Patent Office, London
2. Mishra RS, Ma ZY (2005) Friction stir welding and processing. Mater Sci Eng R 50:1–78
3. Rajakumar S, Muralidharan C, Balasubramanian V (2011) Influence of friction stir welding process and tool parameters on strength properties of AA7075-T6 aluminium alloy joints. Mater Des 32:535–549
4. Sato YS, Onuma T, Ikeda K, Kokawa H (2016) Experimental verification of heat input during friction stir welding of Al alloy 5083. Sci Technol Weld Joining 21(4):325–330
5. Cole EG, Fehrenbacher A, Duffie NA, Zinn MR, Pfefferkorn FE, Ferrier NJ (2013) Weld temperature effects during friction stir welding of dissimilar aluminum alloys 6061–T6 and 7075–T6. Int J Adv Manuf Technol 71(1–4):643–652
6. Nisitani H (1975) Stress intensity factor for the tension of a semi-infinite plate having an oblique or a bent edge crack. Trans Jpn Soc Mech Eng 41(344):1103–1111
7. Y. Murakami ed (1987) Stress Intensity Factors Handbook. Soc Mater Sci Japan 1. McGraw-Hill, Kyoto
8. Sih GC (1974) Strain energy density factor applied to mixed mode crack problems. Int J Fract 10(3):305–321
9. Erdogan F, Sih GC (1963) On the crack extension in plates under plane loading and transverse shear. J Basic Eng 85(4):519–527
10. Oosterkamp O, Oosterkamp LD, Nordeide A (2004) 'Kissing bond' phenomena in solid-state welds of aluminum alloys. Welding J 83:225–231
11. Sato YS, Takauchi H, Park SHC, Kokawa H (2005) Characteristics of the kissing-bond in friction stir welded Al alloy 1050. Mater Sci Eng, A 405:333–338

Effect of Diffusion on Intermetallics at Interface During Friction Stir Welding of Stainless Steel and Pure Titanium

Nikhil Gotawala and Amber Shrivastava

Abstract Relatively lower temperatures during friction stir welding limit the formation of intermetallic compounds while joining dissimilar materials. The objective of this work is to understand the effect of diffusion on intermetallic compound during friction stir welding of stainless steel and pure Ti. Numerical modelling is done to predict temperature distribution and history in the stir region. The temperature history is used to estimate the elemental concentration profiles at the interface. The diffusion of Ti into steel and Fe, and Cr and Ti from steel to titanium leads to the formation of FeTi at the interface. Further β-Ti and α-Ti needles form on the titanium side of the interface, as presence of Fe upon diffusion stabilizes β-Ti. Numerically predicted elemental concentration profiles are compared against the experimental results from a previous study. Numerical results well capture the experimental observations.

Keywords Friction stir welding · Dissimilar joining · Intermetallic compounds · Interfacial diffusion

Introduction

The components made of titanium (Ti) and stainless steel (SS) are assembled in many industries like nuclear and petrochemical [1]. As joining of these materials with fusion welding is difficult due to differences in their thermal expansion coefficients and formation of large amount of intermetallic compounds (IMC), it leads to the formation of cracks [2]. For example, large amount of IMCs like FeTi, Fe_2Ti, and $Ti_5Fe_{17}Cr_5$ have been observed in the laser-welded SS and Ti alloy joints [3]. To hinder Fe and Ti mixing during welding, copper-based filler metal [2] or interlayers [4] were tried with electron beam welding. The copper filler metals and interlayers improved the crack resistance by replacing Fe–Ti IMCs with Ti–Cu IMCs. Solid-state joining is a more efficient way to join dissimilar material due to lower temperatures compared to fusion welding. During fusion welding, materials melt and get mixed all

N. Gotawala · A. Shrivastava (✉)
Department of Mechanical Engineering, Indian Institute of Technology Bombay, Mumbai 400076, India
e-mail: ashrivastava.me@iitb.ac.in

Y. Hovanski et al. (eds.), *Friction Stir Welding and Processing XI*, The Minerals, Metals & Materials Series, https://doi.org/10.1007/978-3-030-65265-4_13

over the fusion zone, which allow the formation of IMCs everywhere in the fusion zone [2]. However, during solid-state joining, IMCs form at the interface of both the materials. Investigators have studied the diffusion bonding SS and Ti alloys, and observed Fe-Ti, Fe–Cr, Cr–Ti, etc., IMCs in the bonded joints [5, 6]. Ni [7], Cu [8], and Ag [9] interlayer between SS and Ti lead to the formation of less brittle IMCs upon diffusion bonding. Diffusion bonding of dissimilar materials leads to relatively larger thickness of IMCs, due to higher incubation time. Incubation time during friction stir welding (FSW) is significantly low. Many combinations of dissimilar materials like Al6061-Copper [10], Al6013-SS [11], and Al6061-Ti6Al4V [12] have been friction stir welded previously. FSW of SS and Ti is also explored by a few studies [13, 14]. FeTi IMC particles and β-Ti were reported in the friction stir welded joints of SS and Ti. Additionally, σ phase of steel and Ni–Ti IMCs was observed by Ishida et al. in SS–Ti joints upon FSW with lap configuration [15].

Many researchers have developed thermo-mechanical models for FSW, to capture temperature, strain rate, and material flow in the stir region. Energy balance can be solved with finite difference approach to predict the temperatures during FSW [16]. Further, residual stresses arising from temperature transients [17] and velocity distribution [18] during FSW have been captured by Finite element method (FEM)-based solvers [17, 18]. The formation of voids in stir region during FSW is also simulated with FEM [19]. Similarly, computational fluid dynamics (CFD)-based approaches have been also deployed to capture the velocity distribution [20] and resulting banded structure [21] in the stir zone. These early works focused on FSW of similar materials. Lately, numerical models have also developed dissimilar FSW. These works primarily focus on the material movement during dissimilar FSW for different material combinations: Al6061-Al5083 [23], Al1100-Al5056 [25], Al6061-trip steel [22, 24], Al5083-SS304 [26], etc. The inter-diffusion of elements during FSW of dissimilar materials leads to the formation of IMCs, which significantly affect the resulting joint properties. Authors previously developed a numerical model to capture diffusion during FSW of Al1050 and Copper [27]. However, the effect of diffusion towards IMCs and other phase transformations was not analysed. The objective of this work is to numerically capture the diffusion at the interface of SS and Ti during dissimilar FSW. This effort will enhance the understanding of role of diffusion on IMC during dissimilar FSW.

Numerical Methodology

A computational fluid dynamics (CFD)-based model is developed to capture the thermal history during FSW of SS and Ti. Further, Fick's second law is used to predict the diffusion at the SS–Ti interface.

Thermo-Mechanical Model

Figure 1 shows the schematic diagram of dissimilar SS–Ti FSW process. The velocity distribution in the stir region is determined by solving conservation of momentum and conservation of mass. The momentum equations are as follows:

$$\rho\left(\frac{\partial u}{\partial t}+u\frac{\partial u}{\partial x}+v\frac{\partial u}{\partial y}+w\frac{\partial u}{\partial z}\right)=-\frac{\partial p}{\partial x}+\frac{\partial \tau_{xx}}{\partial x}+\frac{\partial \tau_{xy}}{\partial y}+\frac{\partial \tau_{xz}}{\partial z} \tag{1}$$

$$\rho\left(\frac{\partial v}{\partial t}+u\frac{\partial v}{\partial x}+v\frac{\partial v}{\partial y}+w\frac{\partial v}{\partial z}\right)=-\frac{\partial p}{\partial y}+\frac{\partial \tau_{xy}}{\partial x}+\frac{\partial \tau_{yy}}{\partial y}+\frac{\partial \tau_{yz}}{\partial z} \tag{2}$$

$$\rho\left(\frac{\partial w}{\partial t}+u\frac{\partial w}{\partial x}+v\frac{\partial w}{\partial y}+w\frac{\partial w}{\partial z}\right)=-\frac{\partial p}{\partial z}+\frac{\partial \tau_{xz}}{\partial x}+\frac{\partial \tau_{yz}}{\partial y}+\frac{\partial \tau_{zz}}{\partial z} \tag{3}$$

Here, u, v, and w are the velocity components in x-, y-, and z-directions (Fig. 1), respectively. τ_{ij} is the shear stress tensor, ρ is the density of material, which is calculated using weighted average method to account for material mixing during FSW, as shown below:

$$\rho = \alpha\rho_{SS} + (1-\alpha)\rho_{Ti} \tag{4}$$

Here, ρ_{SS} and ρ_{Ti} are density of SS and Ti.α is the volume fraction of SS. The equation used for conservation of mass is given below:

$$\frac{\partial u}{\partial x}+\frac{\partial v}{\partial y}+\frac{\partial w}{\partial z}=0 \tag{5}$$

Velocity boundary conditions are applied at FS tool shoulder and material interface, and FS tool pin and material interface. Same are given below:

$$u = (1-\delta)(\omega r cos\theta - V) \tag{6}$$

$$v = (1-\delta)\omega r sin\theta \tag{7}$$

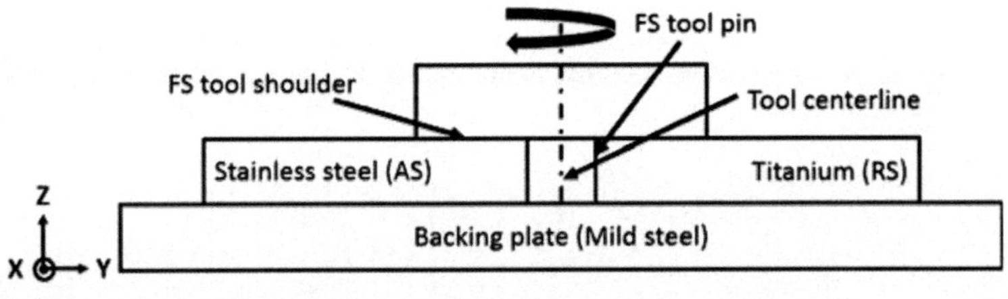

Fig. 1 Schematic diagram of FSW of stainless steel and titanium

$$w = 0 \tag{8}$$

where ω, V, r, and δ are tool rotation speed in rad/sec, feed rate in m/sec, distance from tool center in meters, and slip factor between tool and material surface, respectively. δ is estimated as per [28]

$$\delta = 1 - \exp\left(-\frac{\omega r}{\delta_0 \omega_0 R_s}\right) \tag{9}$$

Here, R_s is the tool shoulder radius, while ω_0 and δ_0 are 400 rpm and 4, respectively [28].

The strain rate tensor, shear strain tensor, and shear stress tensor are calculated by Eqs. 10, 11, and 12, respectively.

$$\dot{\varepsilon}_{ij} = \frac{1}{2}\left(\frac{\partial u_i}{\partial x_j} + \frac{\partial u_j}{\partial x_i}\right) \tag{10}$$

$$\dot{\gamma}_{ij} = 2\dot{\varepsilon}_{ij} \tag{11}$$

$$\tau_{ij} = \mu \dot{\gamma}_{ij} \tag{12}$$

μ is the dynamic viscosity during FSW, which is estimated by

$$\mu = \frac{\sigma'}{3\dot{\varepsilon}_e} \tag{13}$$

Here, σ' is calculated by

$$\sigma' = \alpha \sigma_e^{SS} + (1 - \alpha)\sigma_e^{Ti} \tag{14}$$

σ_e^{SS} and σ_e^{Ti} are effective stresses for SS and Ti and $\dot{\varepsilon}_e$ is the effective strain rate,

$$\dot{\varepsilon}_e = \sqrt{\frac{2}{3}\varepsilon_{ij}\varepsilon_{ij}} \tag{15}$$

The effective stresses for SS and Ti are calculated as per Sheppard and Wright's model (Eq. 16)

$$\sigma_e = \frac{1}{\alpha_e} sinh^{-1}\left[\left(\frac{Z}{A}\right)^{\frac{1}{n}}\right] \tag{16}$$

Z in Eq. 16 is given by

Table 1 Material parameters of Sheppard and Wright's model [29, 30]

Material parameter	Stainless steel	Titanium
$\ln A(s^{-1})$	39.19	34.59
$\alpha_e(Mpa^{-1})$	0.0105	0.0194
Q(J/mol)	475,000	339,055
n	5.055	3.17

$$Z = \dot{\varepsilon}_e \exp\left(\frac{Q}{RT}\right) \quad (17)$$

where Q is the activation energy of material. α_e, A, and n are material-dependent constants. Table 1 lists the values of these constants for SS and Ti.

The steady-state energy balance (Eq. 18) is solved to estimate the temperature distribution during FSW of SS and Ti.

$$k\left(\frac{\partial^2 T}{\partial x^2} + \frac{\partial^2 T}{\partial y^2} + \frac{\partial^2 T}{\partial z^2}\right) + \frac{q_g}{\rho c} + \rho c u \frac{\partial T}{\partial x} + \rho c v \frac{\partial T}{\partial y} + \rho c w \frac{\partial T}{\partial z} = 0 \quad (18)$$

$$q_g = 0.9\tilde{\sigma}\ \bar{\dot{\varepsilon}} \ (19)$$

where *k, c*, ρ, and q_g are the thermal conductivity, specific heat, density, and heat dissipated due to plastic deformation. *k* and *c* are estimated as

$$k = \alpha k_{SS} + (1 - \alpha) k_{Ti} \quad (20)$$

$$c = \alpha c_{SS} + (1 - \alpha) c_{Ti} \quad (21)$$

Table 2 lists the material properties of SS and Ti.

Table 2 Material properties of stainless steel and titanium

Material	Stainless steel	Titanium	Tool
Thermal conductivity (W/mK)	14	16.4	110
Specific heat (J/KgK)	490	523	280
Density (kg/m^3)	8000	4506	19,400
Thermal diffusivity (m^2/s)	3.57×10^{-6}	6.95×10^{-6}	2.02×10^{-5}

Heat flux at tool and workpiece interface due to plastic deformation is estimated as follows:

$$q'' = 0.9\eta(1-\delta)\left(\frac{\sigma_e}{\sqrt{3}}\right)(\omega r \sin\theta - V) \quad (22)$$

where η is fraction of heat flow into the material and given by the ratio of the thermal diffusivities of the workpiece and tool materials. Here, η is fraction of heat flow in to the material and calculated by following equation:

$$\eta = \frac{\beta_{mixed}}{\beta_{tool} + \beta_{mixed}} \quad (23)$$

where β_{mixed} and β_{tool} are thermal diffusivities of mixed material (at tool and workpiece interface) and thermal diffusivity of tool.

The heat transfer coefficient of 5 W/m^2K and surrounding temperature of 298 K are considered at all the exposed surfaces. The heat transfer coefficient at the bottom surface of the workpiece is defined as a function of (Eq. 24).

$$h = 0.003(T - T_{amb})^2 \quad (24)$$

Material Movement Model

Volume of fluid method was used to track fluid interfaces in fluid flow problems. In this analysis, volume of fluid method is used to capture material movement during FSW of SS and Ti. The volume fraction of SS is calculated by

$$\frac{\partial \alpha}{\partial t} + \frac{\partial \alpha u}{\partial x} + \frac{\partial \alpha v}{\partial y} + \frac{\partial \alpha w}{\partial z} = 0 \quad (25)$$

Here, α is the volume fraction of SS and boundary condition on all the workpiece surfaces is given below (s is normal to the surface):

$$\frac{\partial \alpha}{\partial s} = 0 \quad (26)$$

Diffusion Model

Fick's law of diffusion was used to capture the diffusion between two materials,

Table 3 Inter-diffusion coefficients [31]

D_{Fe}^{Ti}	$1.2 \times 10^{-8}\exp(-\frac{110000}{RT})$
D_{Cr}^{Ti}	$0.01 \times 10^{-4}\exp\left(-\frac{158000}{RT}\right)$
D_{Ni}^{Ti}	$0.0092 \times 10^{-4}\exp(-\frac{123900}{RT})$
D_{Ti}^{Fe}	$0.21\exp(-\frac{293200}{RT})$

$$\frac{\partial C}{\partial t} = \frac{\partial}{\partial y}\left(D\frac{\partial C}{\partial y}\right) \tag{27}$$

Here, C is concentration of the element and D is inter-diffusion coefficient of the element. Initially, concentrations of Fe, Cr, and Ni at steel side are 74%, 18%, and 8% and concentration of titanium is 100% at titanium side. Table 3 shows the inter-diffusion coefficients of all the elements.

Previously, authors experimentally investigated the microstructural evolution upon dissimilar FSW of SS and Ti [32]. FSW of SS and Ti was performed at tool rotational speed of 700 rpm and feed rate of 50 mm/min with no tool offset. The inter-diffusion of different elements was analysed by performing the energy-dispersive spectroscopy (EDS) line scan which was performed at the SS–Ti interface. The numerical results from the current study are compared against the experimental results from [32].

Results and Discussion

The predicted temperature distribution during FSW of SS and Ti at the top and cross section is shown in Figs. 2 and 3, respectively. As Fig. 2 shows, the temperature gradient is higher at the leading edge compared to the trailing edge. This is due to relatively lower temperature of incoming material at leading edge and higher temperature of the outgoing material temperature from the stir region. There is asymmetric temperature distribution with respect to transverse axis (x-axis). As Fig. 3 shows, there is also asymmetric temperature distribution in stir region with respect to tool centreline. In stir region, temperature at stainless steel side is higher compared to titanium side. Since the activation energy of stainless steel is higher than the activation energy of titanium (Table 1), the effective stress at stainless steel side is also higher than the effective stress at the titanium side (Eq. 16), for same strain rate and temperature. The higher effective stress at stainless steel side leads to more heat generation and higher temperatures as compared to the titanium side. However, beyond certain distance from the stir region, the temperature at titanium side is higher than the stainless steel side. This is due to higher thermal diffusivity of titanium as compared to stainless steel (Table 2). The temperature is higher at the top surface due to the effect of shoulder and decreased towards the bottom (Fig. 3).

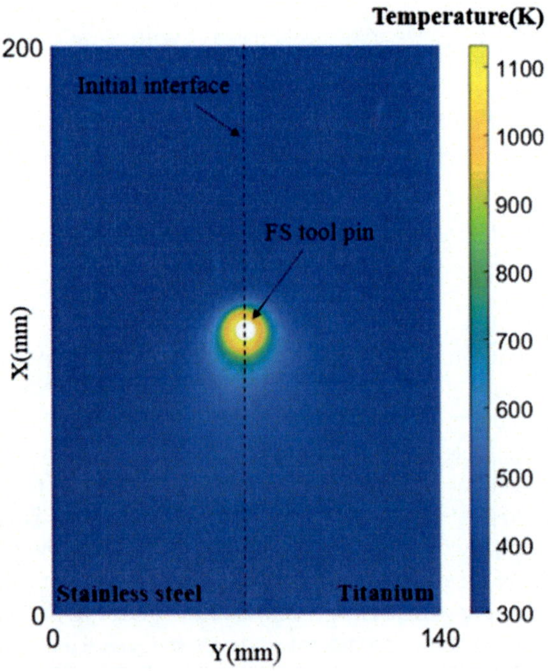

Fig. 2 Predicted temperatures at top surface during FSW of SS and Ti. (Color figure online)

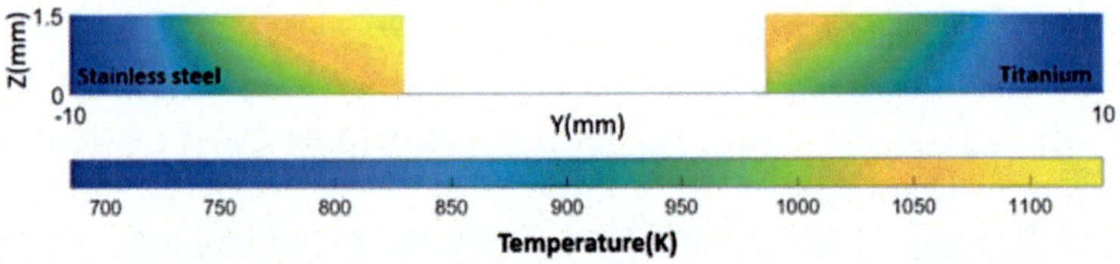

Fig. 3 Predicted temperatures at cross section during FSW of SS and Ti. (Color figure online)

The predicted temperature history was used to determine the diffusion coefficients at each time step. Figure 4 shows the temperature variation with time and corresponding changes in diffusion coefficients, at the SS-Ti interface. The rise in temperature promotes the elemental diffusion across the interface, as all of the inter-diffusion coefficients also increase. These time/temperature-dependent diffusion coefficients along with Fick's law were utilized to estimate the elemental concentration profiles. Figure 5 shows the predicted concentration profiles of Fe, Cr, Ni, and Ti in the vicinity of the SS-Ti interface. It can be noticed that Fe, Cr, and Ni from steel side diffuse to larger distance into the Ti side, as compared to the distance to which Ti diffuses into the steel side. Fe and Ni diffuse farthest from the SS-Ti interface into the Ti side. This is due to the large concentration gradient of Fe across SS-Ti interface and high diffusion coefficient of Ni in Ti, D_{Ni}^{Ti} (Fig. 4). Interestingly, Ti diffuses

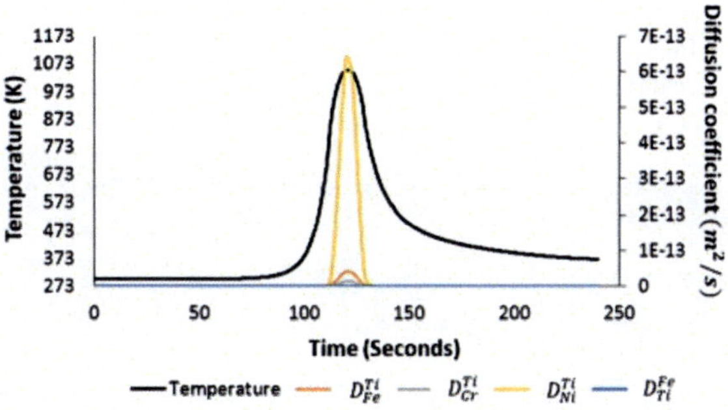

Fig. 4 Predicted temperature history and diffusion coefficients with respect to time at SS-Ti interface. (Color figure online)

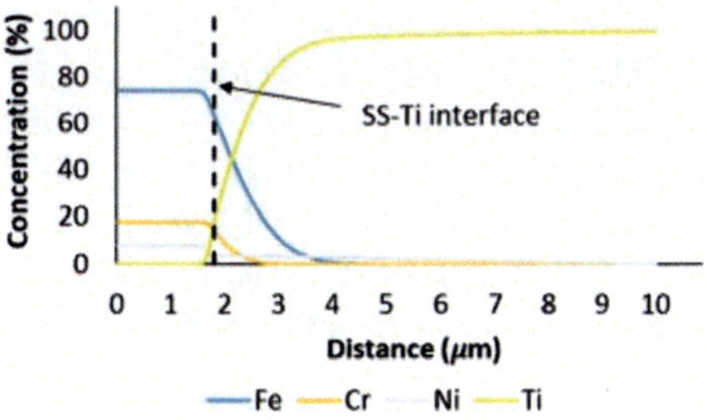

Fig. 5 Predicted concentration distribution across SS–Ti interface. (Color figure online)

to a very limited distance into the SS side (from SS-Ti interface) despite the large concentration gradient across the SS-Ti interface. This is due to very small diffusion coefficient of Ti in Fe, D_{Ti}^{Fe} during the process. The maximum predicted value of D_{Ti}^{Fe} is 5.69×10^{-16} m^2/s, which is three orders smaller than the maximum predicted value of D_{Ni}^{Ti} (=6.43×10^{-13}m^2/s). The maximum predicted values of D_{Cr}^{Ti} and D_{Fe}^{Ti} are 1.41×10^{-14} m^2/s and 4.11×10^{-14} m^2/s, respectively.

Figure 6 shows the micrograph of the SS–Ti interface and nearby region from dissimilar FSW performed at 700 rpm and 50 mm/min. Figure 7 shows the elemental concentration profile across the SS–Ti interface, from EDS line scan (scanned length shown in Fig. 6). The predicted elemental concentration profile (Fig. 5) compares

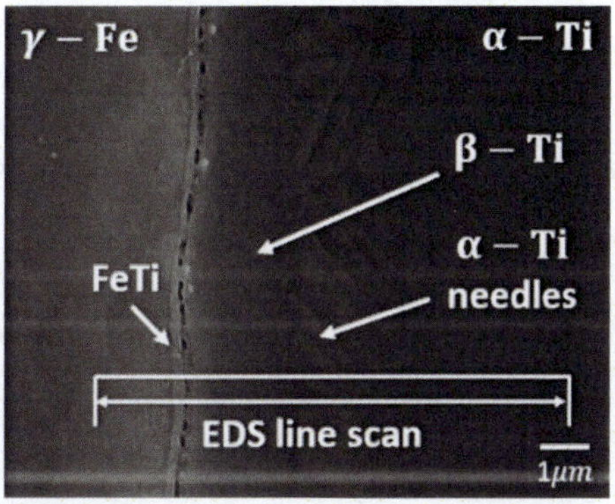

Fig. 6 SEM image of SS–Ti interface

Fig. 7 EDS line scan across S–Ti interface. (Color figure online)

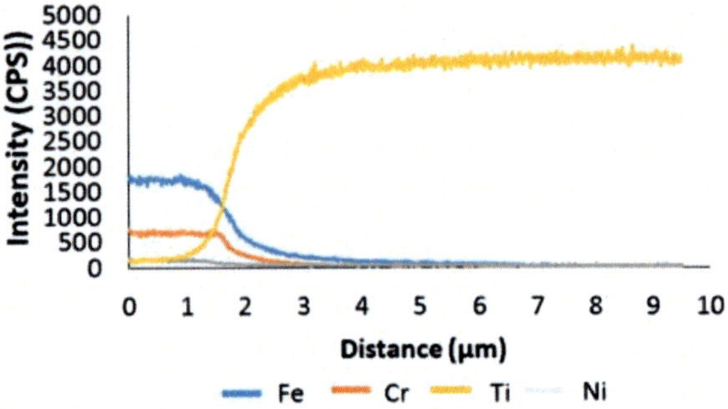

well against the elemental distribution observed experimentally (Fig. 7). From Fig. 6, FeTi, β − Ti, and α − Ti needles can be noticed at and nearby the SS–Ti interface. FeTi can be noticed at the SS–Ti interface, where sufficient concentrations of Fe and Ti are present. Immediately next to SS–Ti interface on Ti side, β − Ti can be noticed (Fig. 6), as the same is stabilized by the presence of Fe and Cr upon diffusion across the interface. Further away from SS–Ti interface into the Ti side, α-Ti needles show up as the Fe and Cr concentration is significantly lowered.

Conclusions

The current work presents a preliminary model to predict the elemental diffusion across dissimilar interface during FSW of SS and Ti. The numerical model accounts for the variation in diffusion coefficients with respect to temperature during the process. The mode predicts a very limited diffusion of Ti into SS, despite the large concentration gradient of Ti across the SS-Ti interface. This is due to very small diffusion coefficient of Ti in Fe. Fe, Cr, and Ni diffuse relatively more into the SS side at the SS–Ti interface. Fe and Ni diffuse farthest from the SS-Ti interface into the Ti side. This is due to the large concentration gradient of Fe across SS-Ti interface and high diffusion coefficient of Ni in Ti. The predicted elemental concentration profile is compared against the experimentally captured elemental distribution across the SS–Ti interface of FSWed joint. The numerical results compare well against the experimental observations. Upon diffusion during FSW, the presence of Fe and Cr in Ti side stabilizes the β − Ti in the immediate vicinity of the SS–Ti interface. Further away from SS–Ti interface into the Ti side, α − Ti needles form as the concentration of Fe and Cr reduces. FeTi IMCs are only noticed at the SS–Ti interface due to the presence of Fe and Ti in sufficient concentrations.

Acknowledgements The authors gratefully acknowledge the partial support of this work by the Science and Engineering Research Board, Department of Science & Technology, Government of India (File no. ECR/2017/000727/ES) and Department of Mechanical Engineering at Indian Institute of Technology Bombay, Mumbai.

References

1. Mudali UK, Rao BA, Shanmugam K, Natarajan R, Raj B (2003) Corrosion and microstructural aspects of dissimilar joints of titanium and type 304L stainless steel. J Nucl Mater 321(1):40–48
2. Wang T, Zhang B, Feng J, Tang Q (2012) Effect of a copper filler metal on the microstructure and mechanical properties of electron beam welded titanium–stainless steel joint. Mater Charact 73:104–113
3. Chen S, Zhang M, Huang J, Cui C, Zhang H, Zhao X (2014) Microstructures and mechanical property of laser butt welding of titanium alloy to stainless steel. Mater Des 53:504–511
4. Tomashchuk I, Sallamand P, Belyavina N, Pilloz M (2013) Evolution of microstructures and mechanical properties during dissimilar electron beam welding of titanium alloy to stainless steel via copper interlayer. Mater Sci Eng A 585:114–122
5. Ghosh M, Bhanumurthy K, Kale GB, Krishnan J, Chatterjee S (2003) Diffusion bonding of titanium to 304 stainless steel. J Nucl Mater 322(2–3):235–241
6. Yuan XJ, Sheng GM, Qin B, Huang WZ, Zhou B (2008) Impulse pressuring diffusion bonding of titanium alloy to stainless steel. Mater Charact 59(7):930–936
7. Kundu S, Chatterjee S (2008) Characterization of diffusion bonded joint between titanium and 304 stainless steel using a Ni interlayer. Mater Charact 59(5):631–637
8. Elrefaey A, Tillmann W (2009) Solid state diffusion bonding of titanium to steel using a copper base alloy as interlayer. J Mater Process Technol 209(5):2746–2752
9. Atasoy E, Kahraman N (2008) Diffusion bonding of commercially pure titanium to low carbon steel using a silver interlayer. Mater Charact 59(10):1481–1490
10. Murr LE, Li Y, Flores RD, Trillo EA, McClure JC (1998) Intercalation vortices and related microstructural features in the friction-stir welding of dissimilar metals. Mater Res Innov 2(3): 150–163
11. Uzun, Huseyin, Claudio Dalle Donne, Alberto Argagnotto, Tommas Ghidini, and Carla Gambaro. (2005) Friction stir welding of dissimilar Al 6013-T4 to X5CrNi18–10 stainless steel." Mater Design 26(1): 41–46
12. Bang K-S, Lee K-J, Bang H-S, Bang H-S (2011) Interfacial microstructure and mechanical properties of dissimilar friction stir welds between 6061–T6 aluminum and Ti-6% Al-4% V alloys. Mater Trans 52(5):974–978
13. Liao J, Yamamoto N, Liu H, Nakata K (2010) Microstructure at friction stir lap joint interface of pure titanium and steel. Mater Lett 64(21):2317–2320
14. Gao Y, Nakata K, Nagatsuka K, Liu FC, Liao J (2015) Interface microstructural control by probe length adjustment in friction stir welding of titanium and steel lap joint. Mater Des 1980–2015(65):17–23
15. Ishida K, Gao Y, Nagatsuka K, Takahashi M, Nakata K (2015) Microstructures and mechanical properties of friction stir welded lap joints of commercially pure titanium and 304 stainless steel. J Alloy Compd 630:172–177
16. Song M, Kovacevic R (2003) Thermal modeling of friction stir welding in a moving coordinate system and its validation. Int J Mach Tools Manuf 43(6):605–615
17. Soundararajan V, Zekovic S, Kovacevic R (2005) Thermo-mechanical model with adaptive boundary conditions for friction stir welding of Al 6061. Int J Mach Tools Manuf 45(14):1577–1587
18. Zhang Z, Chen JT, Zhang ZW, Zhang HW (2011) Coupled thermo-mechanical model based comparison of friction stir welding processes of AA2024-T3 in different thicknesses. J Mater Sci 46(17):5815
19. Al-Badour F, Merah N, Shuaib A, Bazoune A (2013) Coupled Eulerian Lagrangian finite element modeling of friction stir welding processes. J Mater Process Technol 213(8):1433–1439
20. Colegrove PA, Shercliff HR (2005) 3-Dimensional CFD modelling of flow XE "Flow" round a threaded friction stir welding tool profile. J Mater Process Technol 169(2):320–327
21. Tongne A, Desrayaud C, Jahazi M, Feulvarch E (2017) On material flow in friction stir welded Al alloys. J Mater Process Technol 239:284–296

22. Liu X, Lan S, Ni J (2015) Thermal mechanical modeling of the plunge stage during friction-stir welding of dissimilar Al 6061 to TRIP 780 steel. J Manuf Sci Eng 137(5):051017
23. Al-Badour F, Merah N, Shuaib A, Bazoune A (2014) Thermo-mechanical finite element model of friction stir welding of dissimilar alloys. Int J Ad Manuf Technol 72(5–8):607–617
24. Liu X, Chen G, Ni J, Feng Z (2017) Computational fluid dynamics modeling on steady-state friction stir welding of aluminum alloy 6061 to TRIP steel. J Manuf Sci Eng 139(5):051004
25. Yoshikawa G, Miyasaka F, Hirata Y, Katayama Y, Fuse T (2012) Development of numerical simulation model for FSW employing particle method. Sci Technol Weld Joining 17(4):255–263
26. Sadeghian B, Taherizadeh A, Atapour M (2018) Simulation of weld morphology during friction stir welding of aluminum-stainless steel joint. J Mater Process Technol 259:96–108
27. Gotawala N, Shrivastava A (2019) Investigation of interfacial diffusion during dissimilar friction stir welding. In: Friction stir welding and processing X , pp 109–119
28. Nandan RGGR, Roy GG, Lienert TJ, Debroy T (2007) Three-dimensional heat and material flow during friction stir welding of mild steel. Acta Mater 55(3):883–895
29. Marchattiwar A, Sarkar A, Chakravartty JK, Kashyap BP (2013) Dynamic recrystallization during hot deformation of 304 austenitic stainless steel. J Mater Eng Perform 22(8):2168–2175
30. Sajadifar SV, Yapici GG (2014) Workability characteristics and mechanical behavior modeling of severely deformed pure titanium at high temperatures. Mater Des 53:749–757
31. Neumann G, Tuijn C (2011) Self-diffusion and impurity diffusion in pure metals: handbook of experimental data. Elsevier
32. Gotawala N, Shrivastava A (2020) Microstructural analysis and mechanical behavior of SS 304 and titanium joint from friction stir butt welding. Mater Sci Eng A 139658
33. Wang S, Wang K, Chen G, Li Z, Qin Z, Lu X, Li C (2017) Thermodynamic modeling of Ti-Fe-Cr ternary system. Calphad 56:160–168
34. Foul A, Aranas C Jr, Guo B, Jonas JJ (2018) Dynamic transformation of $\alpha \rightarrow \beta$ titanium at temperatures below the β-transus in commercially pure titanium. Mater Sci Eng A 722:156–159

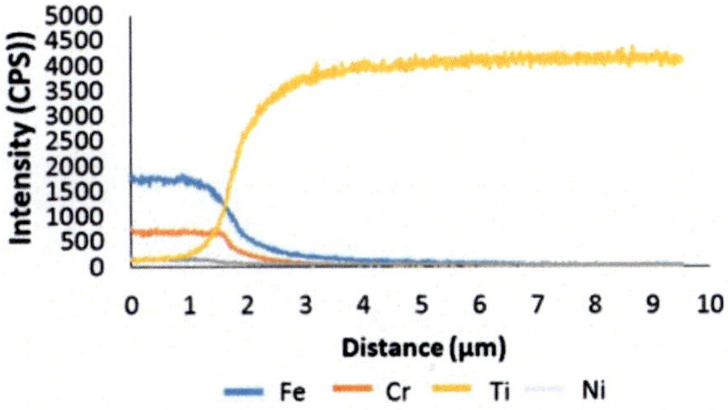

Fig. 7 EDS line scan across S–Ti interface. (Color figure online)

well against the elemental distribution observed experimentally (Fig. 7). From Fig. 6, FeTi, β − Ti, and α − Ti needles can be noticed at and nearby the SS–Ti interface. FeTi can be noticed at the SS–Ti interface, where sufficient concentrations of Fe and Ti are present. Immediately next to SS–Ti interface on Ti side, β − Ti can be noticed (Fig. 6), as the same is stabilized by the presence of Fe and Cr upon diffusion across the interface. Further away from SS–Ti interface into the Ti side, α-Ti needles show up as the Fe and Cr concentration is significantly lowered.

Conclusions

The current work presents a preliminary model to predict the elemental diffusion across dissimilar interface during FSW of SS and Ti. The numerical model accounts for the variation in diffusion coefficients with respect to temperature during the process. The mode predicts a very limited diffusion of Ti into SS, despite the large concentration gradient of Ti across the SS-Ti interface. This is due to very small diffusion coefficient of Ti in Fe. Fe, Cr, and Ni diffuse relatively more into the SS side at the SS–Ti interface. Fe and Ni diffuse farthest from the SS-Ti interface into the Ti side. This is due to the large concentration gradient of Fe across SS-Ti interface and high diffusion coefficient of Ni in Ti. The predicted elemental concentration profile is compared against the experimentally captured elemental distribution across the SS–Ti interface of FSWed joint. The numerical results compare well against the experimental observations. Upon diffusion during FSW, the presence of Fe and Cr in Ti side stabilizes the β − Ti in the immediate vicinity of the SS–Ti interface. Further away from SS–Ti interface into the Ti side, α − Ti needles form as the concentration of Fe and Cr reduces. FeTi IMCs are only noticed at the SS–Ti interface due to the presence of Fe and Ti in sufficient concentrations.

Acknowledgements The authors gratefully acknowledge the partial support of this work by the Science and Engineering Research Board, Department of Science & Technology, Government of India (File no. ECR/2017/000727/ES) and Department of Mechanical Engineering at Indian Institute of Technology Bombay, Mumbai.

Process Robustness of Friction Stir Dovetailing of AA7099 to Steel with In Situ AA6061 Interlayer Linking

Md Reza-E-Rabby, Timothy Roosendaal, Piyush Upadhyay, Nicole Overman, Joshua Silverstein, Martin McDonnell, and Scott Whalen

Abstract Rolled homogeneous armor (RHA) steel was joined to aluminum alloy AA7099-T7451 with an interlayer of AA6061-T651 by single-pass friction stir dovetailing (FSD). The evolution of the AA7099/AA6061 and AA6061/steel interfacial microstructures and mechanical properties of the joints were examined. The AA6061 interlayer prevented Zn migration from the AA7099 to the steel during FSD and inhibited formation of brittle, Zn-rich, Fe–Al intermetallic compounds within the joint. The flow of the dissimilar aluminum alloys at the interface was controlled with a partially threaded friction stir welding pin and a dovetail geometric configuration optimized in previous work. The process robustness was explored to establish parameter windows by elucidating parametric effects on the relationship between process conditions and the joint's microstructure and mechanical properties. Ranges of process forge force (50–70 kN) and interface temperature (475–515 °C) consistently yielded joints with adequate performance. The load-carrying capacity in lap shear tensile tests was 2025 ± 100 N/mm.

Keywords Friction stir dovetailing · FSW · AA7099 · AA6061 · Rolled homogenous armor · RHA · Dissimilar metals

Introduction

Friction stir welding was originated in 1991 by The Welding Institute (TWI) in the UK [1] to solve joining challenges in aluminum alloys. Now, after almost three decades of research, this technology has expanded to joining of dissimilar material systems, including Al to steel in both butt [2] and lap [3] joint configurations. However, the scalability of this process for increasingly thick sections of Al and steel is challenging

M. Reza-E-Rabby (✉) · T. Roosendaal · P. Upadhyay · N. Overman · J. Silverstein · S. Whalen
Energy and Environment Directorate, Pacific Northwest National Laboratory, 902 Battelle Blvd, Richland, WA 99354, USA
e-mail: md.reza-e-rabby@pnnl.gov

M. McDonnell
U.S. Army Combat Capability Development Center – Ground Vehicle Systems Center, 6501 E 11 Mile Road, Warren, MI 48397, USA

Y. Hovanski et al. (eds.), *Friction Stir Welding and Processing XI*, The Minerals, Metals & Materials Series, https://doi.org/10.1007/978-3-030-65265-4_14

in three areas: (1) tooling requirements and costs [4], (2) process robustness (steel does not co-deform with Al at temperatures below the melting temperature of Al), and (3) repeatability (the reaction layer that forms an Fe–Al intermetallic is very sensitive to process parameters and elemental compositions of alloys). The friction stir dovetailing (FSD) [5]—a derivative technology of friction stir welding (FSW)—was developed to overcome these challenges associated with joining materials that have vastly different melting temperatures. FSD eliminates the difficulties associated with conventional FSW of dissimilar metals; the goal is to provide mechanical interlocking through the dovetail as well as metallurgical bonding at the dovetail interface. To do this, a low-cost, customized FSW tool disrupts the dissimilar interface by introducing a local high-temperature deformation and metallurgical reaction. Having successfully joined AA6xxx to rolled homogeneous armor (RHA) steel [6], incorporation of AA7xxx series Al alloy would expand the high-strength applications of lightweight materials. However, very few studies have been performed on joining AA7xxx to steel in the solid-phase materials processing regime. Some examples are joints using FSW [7–9], ultrasonic spot welding [10], and FSD [11]. However, Zn, the principal alloying element in AA7xxx, dissolves rapidly and forms a brittle eutectic Zn–Al at the Al–Fe intermetallic layer [10].

In previous work, nanocrystalline, silicon-rich $FeAl_3$ intermetallic compounds (IMC) were formed in FSD of AA6061 to RHA plates with highly ductile joints [6]. However, FSD of AA7099 to RHA demonstrated brittle failure with less load-carrying capacity than that of the AA6061-RHA FSD joint because an eta prime (η') IMC formed during FSD of AA7099 to RHA [11]. Therefore, this study takes advantage of AA6061 serving as interlayer to form a single-pass FSD joint of AA7099 and RHA that eliminates the formation of Zn-rich IMCs and generates the desirable Si-rich, Fe–Al IMC at the AA6061-RHA interface. The efficacy of the joint was evaluated using lap shear tensile test and scanning electron microscope (SEM) imaging. The repeatability and robustness of this method with regard to process parameters and tool life were also recorded systematically in this study.

Materials and Experimental Procedure

Two types of Al alloys, AA7099-T7451 plates (10.5 mm × 300 mm × 150 mm), AA6061-T651 strips (2 mm × 300 mm × 16 mm), and RHA (specification per MIL-DTL-12560 J) steel plates (12.7 mm × 300 mm × 150 mm) were employed in this study. The chemical compositions of these materials are presented in Table 1. The materials, joint configurations/placement of alloys, tool used in this study, and post-weld cross-sectional macrograph are shown in Fig. 1a–d. The free-floating AA6061 strip was fitted within the pre-machined dovetail groove in the RHA, after which AA7099 was overlaid on RHA and clamped. The narrow AA6061 strip was restricted to movement in the welding direction, as shown in Fig. 1b (restraining object is not shown in the image). In addition, the extrusion of AA7099 into the dovetail groove during the FSD process provided the compressive load on AA6061 in front of the tool

Table 1 Chemical composition of AA7099, AA6061, and RHA plates (wt%)

Alloy	AA7099-T7451	AA6061-T651	RHA
Zn	7.4–8.4	≤0.25	–
Mg	1.6–2.3	0.8–1.2	–
Cu	1.4–2.1	0.15– 0.4	0.25
Mn	≤0.04	≤0.15	1 ± 0.15
Cr	≤0.04	0.04–0.35	1.25 ± 0.15
Zr	0.05–0.15	–	0.1
Si	≤0.12	0.4–0.8	0.6 ~ 1.0
Ti	≤0.06	≤0.15	0.1
C	–	–	0.3 ± 0.05
Ni	–	–	0.1
Mo	–	–	0.2
Fe	≤0.15	≤0.7	Balance
Al	Balance	Balance	0.1

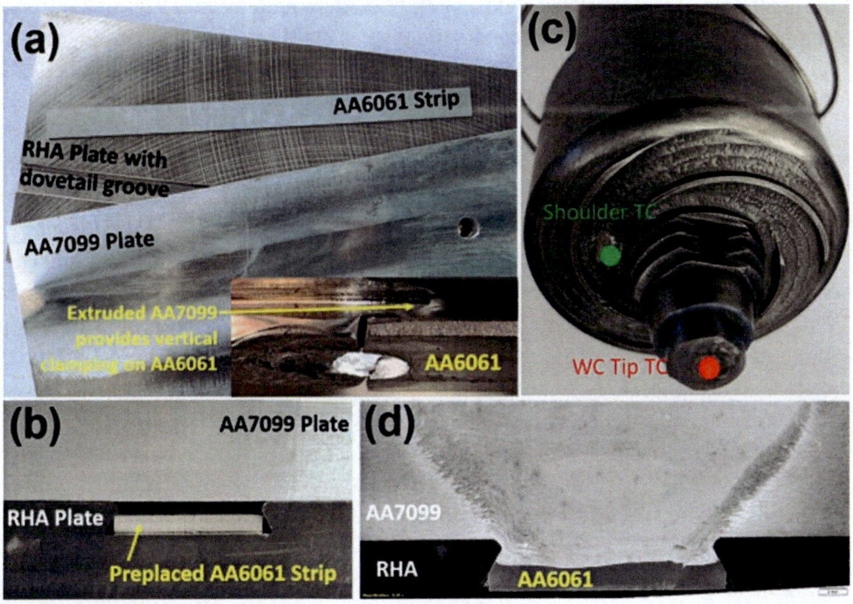

Fig. 1 **a** Materials used in this study: AA7099 plate, RHA plate with dovetail groove, AA6061 strip; **b** placement of material before the FSD process; **c** FSD tool with thermocouple (TC) positions for controlling (red dot) and recording (green dot) the temperature; **d** cross-sectional macrograph of the FSD joint produced in this study. (Color figure online)

Table 2 Summary of welding parameters in single-pass FSD

Trial ID	Forge force (commanded)	Plunge depth into RHA	WC temperature (Controlled)	Rotational speed
	kN	mm	°C	RPM
A	57–71	0.025–0.102	475–485	110–160
B	50–57	0.051–0.127	490–500	180–190
C	48–65	0.025–0.038	490	105–185
D	58–74	0.025–0.051	490	150–180
E	60–75	0.025–0.051	425–430	90–100
F	50–59	0.025–0.051	510	170–210

as it traveled, as illustrated in the inset of Fig. 1a. The customized FSD tool (shown in Fig. 1c was introduced to create the single-pass FSD joint between AA7099 and RHA with an AA6061 interlayer. It should be noted here that the thread was partially machined off the pin tip to minimize the thread-induced vertical material movement during the process as revealed at the AA7099-AA6061 interface in Fig. 1d. Optimized geometric dimensions of the dovetail and the FSD tool selection were adopted based on a previous study [11]. The cross-sectional macrograph in Fig. 1d illustrates the mixing of AA7099 and AA6061 with minimum vertical mixing.

All welds were performed at a constant welding speed of 76.2 mm/min. The temperature at the tip of the tungsten–carbide (WC) insert in the FSD tool was controlled by dynamically modulating spindle torque with a temperature control algorithm [12]. The plunge depth of the FSD tool was controlled with a machine deflection compensation algorithm at different commanded forge forces listed in Table 2. It should be noted here that in each trial, the forge force and WC temperature range were varied to observe the response in other parameters, such as plunge depth and rotational speed, and their effect on the lap shear tensile test. These experiments were also designed to help understand the maximum repeatable force that an FSD tool can endure before breaking.

Each trial weld was sectioned using a water jet to obtain specimens for weld cross-sectional examination and lap shear tensile tests. The average width of samples for lap shear tensile tests was 14 mm and the testing was performed using a 222 kN (50 kip MTS) test frame at a strain rate of 2.54 mm/min. Standard grinding and polishing sequences were followed for metallographic sample preparation, and the final polished surface was obtained using colloidal silica (<0.05 μm). The cross section of the dovetail joint was examined using an Olympus BX51 optical microscope in reflection mode as well as a JEOL 7600 field-emission SEM.

Results and Discussion

Forge Force Consideration on Joint Performance

Figure 2a presents the load-carrying capacity (normalized load, i.e., load per weld length) as a function of forge force during FSD trials at variable interface temperature. It should be noted here that the dark continuous horizontal line in Fig. 2a (1050 N/mm strength level) indicates the minimum load-carrying capacity of the FSD joint contribution solely from the mechanical interlocking of the dovetail. Therefore, this is considered as the baseline for the load-carrying capacity when no intermetallic was formed at the dissimilar interface. Figure 2b clearly shows the undisrupted interface between Al and steel, which can slide over each other after the FSD process because there is no metallurgical bonding. The gray continuous vertical line in Fig. 2a at 48 kN forge force designates the minimum forge force required to form a fully consolidated FSD joint with a disrupted RHA-Al interface at the dovetail. Thus, the functionality of the FSD joint (both mechanical interlocking and metallurgical bonding) will be hindered when applied forge force is below 48 kN. The gray and dark dashed vertical

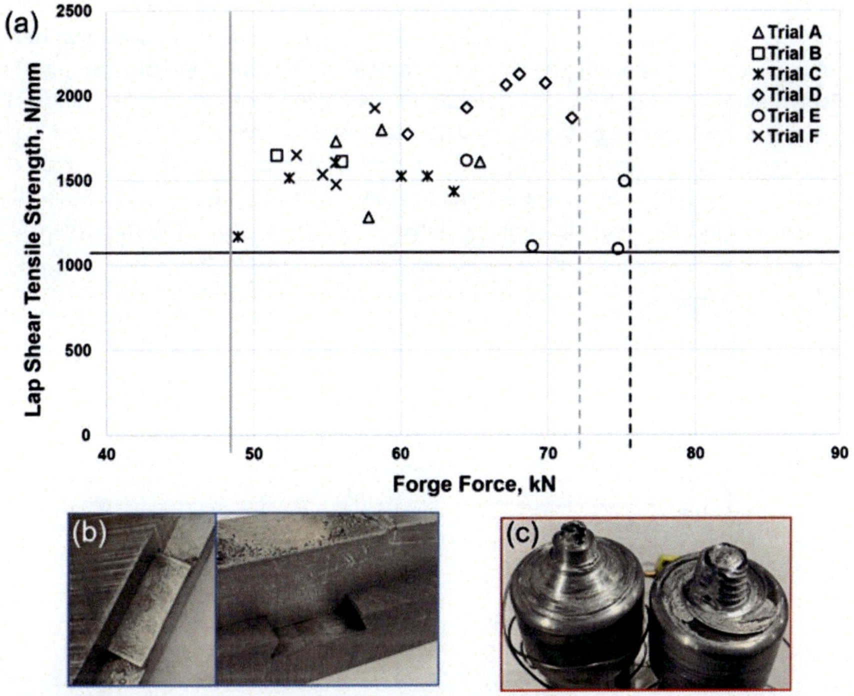

Fig. 2 **a** Lap shear tensile strength as a function of applied forge force during FSD process, **b** example of FSD joint produced solely by mechanical interlocking, and **c** failed FSD tools (tips of WC inserts broken) at process force of 75 kN or more. (Color figure online)

lines are the representative forge force levels of 72 and 75 kN. The process force of 72 kN resulted in an excessively disrupted interface with hook formation (will be described with microstructure in subsequent section) that affects the lap shear tensile strength and jeopardizes the tool life. Moreover, the WC part of the tool often fails when the forge force for FSD is 75 kN and beyond. Figure 2c shows two broken tools immediately after FSD when the forge force reached 75 kN. The different trials in Fig. 2a show that an intermediate forge force leads to higher load-carrying capacity. For instance, at 490 °C interfacial temperature and 68 kN process force (Trial D), the maximum lap shear tensile strength was 2125 N/mm, while at 510 °C and 59 kN process force (Trial F), the maximum lap shear tensile strength was 1925 N/mm.

Effects of AA6061-RHA Interface Temperature on Joint Performance

It is critical to know an optimum AA6061-RHA joining temperature that yields maximum joint performance and suppresses IMC formation [6]. This study aims to understand the range of this critical temperature by adding an AA6061 interlayer in the joint between AA7099 and RHA. Therefore, the FSD joints were made in an interface temperature range of 430–510 °C. Figure 3 shows the lap shear tensile strength as a function of the WC tip temperature, which represents the interface temperature. Similar to Fig. 2, the horizontal dark continuous line in Fig. 3 shows the minimum load-carrying capacity for a dovetail interlock free of IMCs. This study also shows that for an interface temperature below approximately 475 °C (as indicated by the vertical gray line), the interface is weak, so the joint strength is ascribed to the dovetail interlocking alone (at lower forge force, $\leq$ 48 kN) or a vigorously

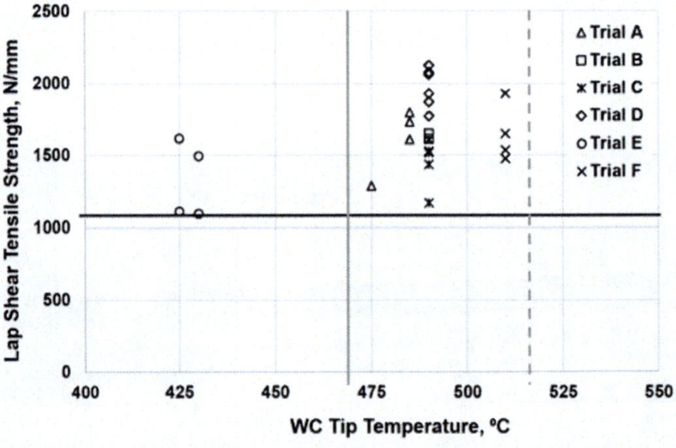

Fig. 3 Lap shear tensile strength as a function of AA6061-RHA interface temperature during the FSD process

disrupted interface (at higher forge force, ≥72 kN). Previous studies have reported that uncontrolled growth of IMCs at high heat input during the FSW process resulted in decreased joint performance [13, 14]; therefore, this study does not exceed the interfacial temperature of 510 °C. In the hybrid FSD joint of AA7099 to RHA with AA6061 interlayer, maximum lap shear joint strength was observed at 2125 N/mm and 490 °C, which supports the previous study [6]. Interestingly, there is a 9% decrease in joint strength at 510 °C compared to the highest load-carrying capacity, which is consistent with the general trend of the temperature–IMC thickness–joint strength correlation [9, 13, 15, 16].

Process Windows and Correlation with Mechanical and Microstructural Analysis

Figure 4 presents dovetail cross sections and interface microstructures for welding with two representative pairs of process parameters: (a) 66 kN forge force and 490 °C interface temperature and (b) 74 kN forge force and 490 °C interface temperature. These two samples were chosen to demonstrate the effect of plunge force and interface temperature (490 °C) on resulting microstructures. The trios of low-resolution SEM images in Fig. 4a and b demonstrate that the degree of interface disruption for lower forge force is less than that for higher force. Moreover, a continuous metallurgical bond was observed in the high-resolution SEM micrographs in Fig. 4a at low

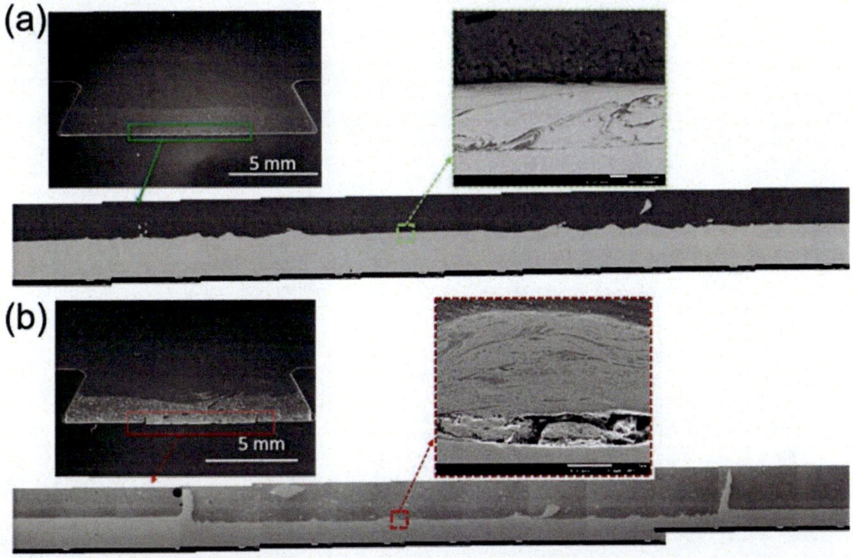

Fig. 4 Representative SEM analyses of FSD joints at **a** 66 kN forge force and 490 °C interface temperature and **b** 74 kN forge force and 490 °C interface temperature. (Color figure online)

process force. However, high process force (74 kN) resulted in a hook formation, as shown in the SEM micrographs in Fig. 4b. Furthermore, the interface was separated, as observed in the high-resolution SEM images in Fig. 4b; this is believed to be IMC that broke after being formed. As reported in Fig. 2-a and Fig. 3, the load-carrying capacities of the samples shown adjacent to these microstructures are 2125 N/mm (Fig. 4a: continuous IMCs) and 1865 N/mm (Fig. 4b: discontinuous IMCs). This 13% difference in load-carrying capacity of FSD joints with identical mechanical interlocking is attributed to differences in the load-transfer mechanisms of continuous IMCs, which provide higher capacity, versus the broken/discontinuous IMCs with hooking. In both cases, the load-carrying capacities in lap shear tensile tests were improved, by 100% and 78%, respectively, for samples with continuous (ref. Fig. 4a) and discontinuous (ref. Fig. 4b) IMCs beyond the load-carrying capacity of the dovetail interlock alone (1050 N/mm).

Figure 5 is a 3-D bubble plot of forge force versus interface temperature versus joint strength in which the bubble sizes are normalized by the base strength (1050 N/mm) that is translated solely through the dovetail mechanical interlock. The lower and upper bounds for forge force (50 kN–70 kN) and temperature (475–515 °C) were established in Fig. 5 as denoted by continuous and dashed lines, respectively. For the forge force bounds, (a) no IMCs are formed at or below 48 kN, and (b) formation of hooks, interfacial disruption/breaking of IMCs, and tool failure occur at or above 72 kN. For the temperature bounds, (a) weak or no IMCs are formed below 475 °C and (b) brittle IMCs are formed at temperatures above 510 °C. Figure 5 also indicates that the forge force requirement for obtaining maximum joint strength dropped with increasing interface temperature as indicated by the shaded ellipse in the figure.

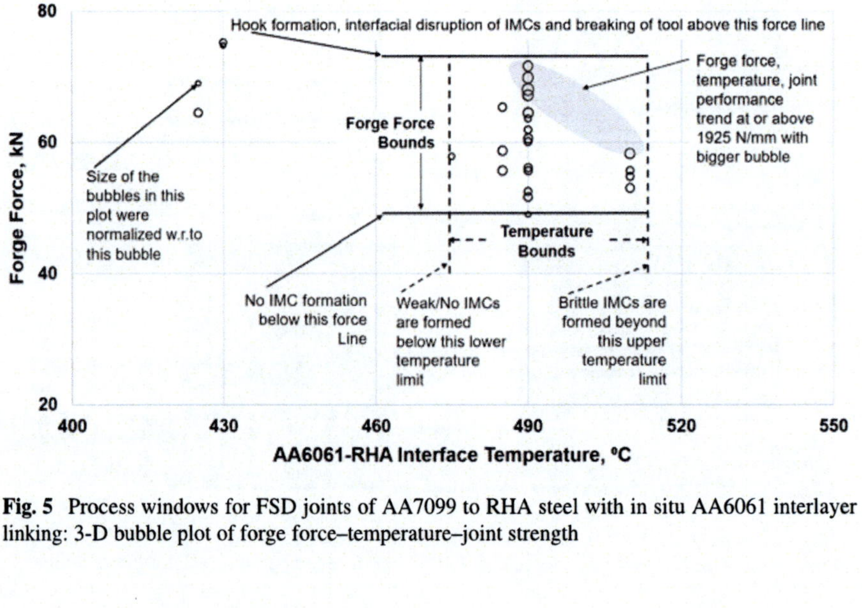

Fig. 5 Process windows for FSD joints of AA7099 to RHA steel with in situ AA6061 interlayer linking: 3-D bubble plot of forge force–temperature–joint strength

Conclusions

The following conclusions can be drawn from this study:

1. A single-pass, hybrid FSD joint was developed for joining AA7099 to RHA steel with an AA6061 interlayer in which simultaneous mixing of AA7099 with AA6061 and desired IMC bonding of AA6061 with RHA were produced.
2. An optimized parameter window was established that correlates force, interface temperature, and joint performance.
3. With the suggested FSD parameters, desirable joint performance can be achieved by avoiding hook formation or premature failure of tool.
4. Under the studied configuration of plates and dovetail geometries, 63 ± 7 kN forge forces and 500 ± 10 °C interface temperature resulted in a load-carrying capacity of 2025 ± 100 N/mm in lap shear tensile tests.

Acknowledgements This work was supported by the U.S. Army Combat Capabilities Development Command Ground Vehicle Systems Center [Military Interdepartmental Purchase Requests 10834736, 11008837, and 11079043]. PNNL is operated by Battelle Memorial Institute for the United States Department of Energy under contract DE-AC05-76RL01830. The authors would also like to thank PNNL staff member Anthony Guzman for helping with microstructural sample preparation for this study.

References

1. Thomas W, Nicholas E, Needham J, Murch M, Templesmith P, Dawes C (1991) GB Patent application no. 9125978.8, International patent application no. PCT/GB92/02203
2. Lee W-B, Schmuecker M, Mercardo UA, Biallas G, Jung S-B (2006) Interfacial reaction in steel–aluminum joints made by friction stir welding. Scripta Mater 55:355–358
3. Kimapong K, Watanabe T (2005) Effect of welding process parameters on mechanical property of FSW lap joint between aluminum alloy and steel. Mater Trans 46:2211–2217
4. Thompson B (2011) Tungsten-based tool material development for the friction stir welding of hard metals. Friction Stir Weld Process V I:105–112
5. Whalen SA, Ross KA, Reza-E-Rabby M, Hovanski Y (2018) System and process for joining dissimilar materials and solid-state interlocking joint with intermetallic interface formed thereby, in, Google Patents
6. Reza-E-Rabby M, Ross K, Overman NR, Olszta MJ, McDonnell M, Whalen SA (2018) Joining thick section aluminum to steel with suppressed FeAl intermetallic formation via friction stir dovetailing. Scripta Mater 148:63–67
7. Borchers T, Seid A, Babu S, Shafer P, Zhang W (2015) Effect of filler metal and post-weld friction stir processing on stress corrosion cracking susceptibility of Al–Zn–Mg arc welds. Sci Technol Weld Joining 20:460–467
8. Goel P, Khan NZ, Khan ZA, Ahmari A, Gangil N, Abidi MH, Siddiquee AN (2019) Investigation on material mixing during FSW of AA7475 to AISI304. Mater Manuf Processes 34:192–200
9. Tanaka T, Morishige T, Hirata T (2009) Comprehensive analysis of joint strength for dissimilar friction stir welds of mild steel to aluminum alloys. Scripta Mater 61:756–759

10. Xu L, Wang L, Chen Y-C, Robson JD, Prangnell PB (2016) Effect of interfacial reaction on the mechanical performance of steel to aluminum dissimilar ultrasonic spot welds. Metallur Mater Trans A 47:334–346
11. Reza-E-Rabby M, Whalen S, Ross K, McDonnell M (2019) Joining AA7099 to Ni-Cr-Mo steel using friction stir dovetailing. In: Friction stir welding and processing X. Springer, pp 171–180
12. Ross K, Reza-E-Rabby M, McDonnell M, Whalen SA (2019) Advances in dissimilar metals joining through temperature control of friction stir welding. MRS Bull 44:613–618
13. Chen C, Kovacevic R (2004) Joining of Al 6061 alloy to AISI 1018 steel by combined effects of fusion and solid state welding. Int J Mach Tools Manuf 44:1205–1214
14. Watanabe T, Takayama H, Yanagisawa A (2006) Joining of aluminum alloy to steel by friction stir welding. J Mater Process Technol 178:342–349
15. Kimapong K, Watanabe T (2005) Lap joint of A5083 aluminum alloy and SS400 steel by friction stir welding. Mater Trans 46:835–841
16. Ogura T, Saito Y, Nishida T, Nishida H, Yoshida T, Omichi N, Fujimoto M, Hirose A (2012) Partitioning evaluation of mechanical properties and the interfacial microstructure in a friction stir welded aluminum alloy/stainless steel lap joint. Scripta Mater 66:531–534

Part VI
Modeling: Process and Properties

The Development of FSW Process Modelling for Use by Process Engineers

M. Lewis and S. D. Smith

Abstract Computational resources are such that process engineers might consider the use of computer models during the development of the manufacturing process of complex fabrications. However, it might not be possible for such an engineer to obtain reliable materials data, because manufacturing processes tend to require property data for a set of loading conditions that is so extensive that such data is only available for a small number of materials. The current paper discusses the development of a Friction Stir Weld (FSW) modelling method using only generally available material data for a wide range of materials. The method assumes that accuracy depends upon the conditions close to the FSW tool plus some straightforward mechanics and thermodynamics remote from the tool. The material model was implemented in STAR-CCM +, a commercially available computational fluid dynamics code, and the results from the model have been compared with measurements made on a test weld.

Keywords Friction stir welding · CFD · Modelling · Flow · Temperature · Mixing · AA2024

Background

Welding techniques are used throughout the manufacturing sector to create complicated part forms from simpler geometries cost-effectively. The selection of the welding process and the subsequent identification of an acceptable welding procedure are largely conducted by experts with considerable experience. Increasingly, however, knowledge can be supplemented by computer-based modelling methods that are able to provide a depth of insight that it would be difficult to achieve solely

M. Lewis (✉)
FTS Engineering Answers Ltd., London, UK
e-mail: mike.lewis@fts-engineeringanswers.com

S. D. Smith
Transforming Stress Ltd., London, UK
e-mail: simon@transformstress.co.uk

Y. Hovanski et al. (eds.), *Friction Stir Welding and Processing XI*, The Minerals, Metals & Materials Series, https://doi.org/10.1007/978-3-030-65265-4_15

from weld coupons and standard weld procedure testing. As a trivial example, it is possible to use a computer model to predict the effect of material strength upon the welding parameters without the need to source coupons with the desired level of variability. Modelling of the melt pool of fusion welds and in additive manufacturing is possible [1, 2], but such simulations require an advanced knowledge of simulation software and simulation techniques as well as a large material properties dataset. Friction Stir Welding (FSW) is an altogether much simpler physical process. The heating is generated by adiabatic shear and heat flow occurs by both material movement (advection) and conduction near the tool and by conduction only remote from the tool. The most complicated aspect of FSW simulation is therefore the development of a material model to predict the relationship between the shearing rate and the shear forces in the vicinity of the tool pin and the tool shoulder. It appears, therefore, that it should be possible to provide FSW process engineers with a relatively straightforward tool for the simulation-based welding trials. An approach is discussed together with a worked example.

It is not usual to adopt a first-principles constitutive representation of material in models of FSW, so analysts usually develop their models based on material behaviour at a range of temperatures from ambient up to near melting and for strain rates from pseudo-static up to values in the region of 1000/sec. This dataset is considerably smaller than the scope of properties needed for simulations of fusion welding, but it is still significant. Colegrove [3], for example, used the Sellars–Tegart equation to give material properties for models of FSW. He therefore required for constants that would not be widely available for materials, including an activation energy. It is therefore of interest to consider the use of an easier to obtain subset of data based upon material data that would be readily available to a FSW process engineer. A possible choice of material data is discussed, and an implementation is presented.

Materials Data for FSW Simulation

The weld procedure specification for a FSW is clearly different from those made for any form of fusion welding. Fusion welding tends to be defined in terms of heat input (in units of energy per unit length of weld made) and travel speed. A FSW procedure will also be specified in terms of travel speed, but the energy needed to make a unit length of weld is not defined. Instead, the other major specified parameter is the tool rotation speed. This is an attractive aspect of the FSW process. The power consumed arises from the torque needed to rotate the spinning tool, so that FSW has a natural feedback loop. A lowering of the temperature of the material around the tool will tend to increase the torque needed to rotate the tool and vice versa. In other words, FSW tends to be a solid-state welding process because the input power would drop considerably if the material around the tool became molten. A schema of this behaviour is shown in Fig. 1.

The built-in feedback mechanism of FSW tends to keep the material near the tool within a small window of strain rates and temperatures. It is really within this window

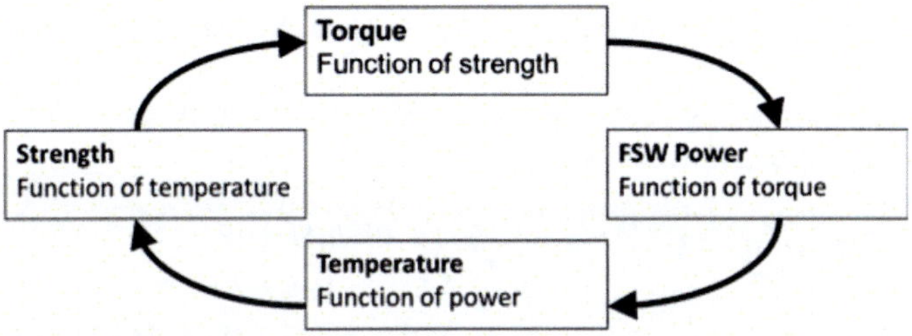

Fig. 1 Simple relationship between the main characteristics that determine the conditions that are generated during friction stir welding

that the material data is needed for a FSW simulation. For most practical conditions of welding, the temperatures will be close to but below the material melting point, and it is the temperature close to the tool that dominates properties of the weld being made. An equilibrium of torque must be developed in every ring around the tool, and the torque in each ring arises from the strain rates combined with the local material resistance to straining. The current material assumption considers that the material behaviour in the heavily straining region is dominated by an exponentially weakening trend of material strength as the material temperature approaches its melting point. This means that the temperature near the tool cannot exceed the melting point because the material resistance would then be too small to develop the torque needed to sustain the high temperature at the tool. A simple relationship between material strength (σ_{0SR}^{T}) at pseudo-static strain rates and temperature, T is assumed as shown below:

$$\sigma_{0SR}^{T} = max\left[\begin{array}{c}\sigma_{0ST}^{RT}\left(1-\left(\frac{T-RT}{T_M-RT}\right)\right)\\ \sigma_{0SR}^{TTL}\left(1+tanh\left\{\frac{T_T-T}{T_M-T_T}\right\}\right)\end{array}\right] \quad (1)$$

The strength is expressed in terms of the room temperature pseudo-static strength, σ_{0ST}^{RT} and the melting temperature T_M. A transition temperature, T_T is defined. The material strength drops linearly with temperature below this point and it drops exponentially at temperatures above this point. The material strength σ_{0SR}^{TTL} is the value determined from the linear portion of the curve at the transition temperature (Eq. 2).

$$\sigma_{0SR}^{TTL} = \sigma_{0ST}^{RT}\left(1-\left(\frac{T_T-RT}{T_M-RT}\right)\right) \quad (2)$$

Equations 1 and 2 provide a material strength characteristic for FSW modelling that can be developed using only the material room temperature strength and the material melting temperature. As discussed, it has the correct characteristics for the loss of strength in the vicinity of the melting point, and, of course, the correct room temperature strength. As proposed, this system can be used for any material to develop a model of FSW as part of an overall welding process development without the need for an advanced dataset or to undertake an extensive and expensive material testing program.

A simple strain rate sensitivity was added as shown in Eq. 3 to augment the pseudo-static strength behaviour.

$$\sigma_{SR}^{T} = \sigma_{0SR}^{T}\left(1 + \frac{0.1\dot{\varepsilon}}{10^{3}}\right) \tag{3}$$

Here, $\dot{\varepsilon}$ is the strain rate of interest.

Software for FSW Simulation

The material model presented in the previous section was developed to provide a FSW process engineer with a straightforward and low-cost method of selecting material data for a FSW model. It is also important, however, that it is simple for the process engineer to be able to create a model that uses this simple material dataset. Early models of FSW (e.g. Askari et al. [4]) used non-commercial software. Now, however, advances in IT systems mean that is straightforward to use commercially available software to simulate FSW. The current work was undertaken using the Computational Fluid Dynamics (CFD) code STAR-CCM + [5] which provides user-friendly and simple-to-learn facilities for the development of a FSW model.

A CFD code adopts a definition of material resistance to shearing in terms of viscosity, μ. It is important, therefore, to convert the equations presented in the previous section into a form that is suitable for CFD. The strength properties presented in the previous section will usually be derived from a tensile test, so the stress, σ_{SR}^{T}, and the strain rate, $\dot{\varepsilon}$, will be direct tensile values. The classical viscosity can be derived from this data via the following equation:

$$\mu = \frac{\sigma_{SR}^{T}}{3\dot{\varepsilon}} \tag{4}$$

Validation Test Weld

The Cambridge University Engineering Department has published details of the experimental investigation of a FSW that is freely available via the Internet [6]. It was originally proposed that the site would provide a catalogue of test measurements made on FSWs. Currently, it contains a description of tests conducted on one combination of plate size, material, and welding procedure. However, the test weld investigated was thoroughly analysed, so that force measurements, temperature measurements, and measurements of material movement (via imaging techniques used to measure

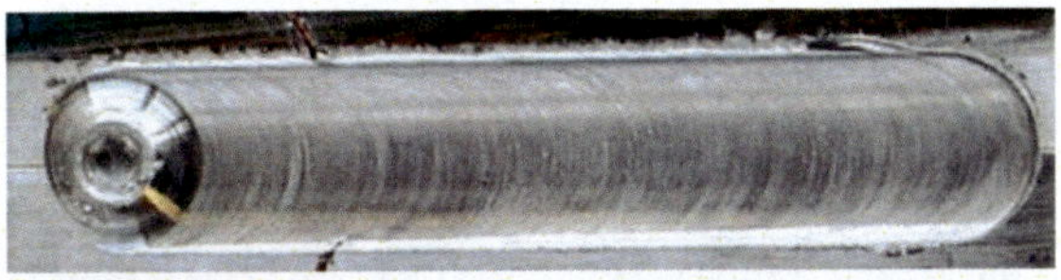

Fig. 2 Surface of FSW test specimen T4. This specimen did not have a flow measurement copper insert. (Color figure online)

the shape of a copper shim after welding inserted into the path of the weld before welding).

The weld was made between 5.85 mm thick plates of aluminium alloy 2024 of temper T3. Each of the two plates being welded together was 150 × 60 mm. Welding was undertaken along the 150 mm side of the plates. The FSW tool had an 18 mm diameter shoulder and a TWI Ltd 5651 pin design with an M6 × 0.8 thread. The pin length was 5.3 mm. Welding was undertaken at a tool rotation speed of 600 RPM and a tool traversing speed of 80 mm/min. Figure 2 shows the top surface of specimen T4 after welding. This was a plain specimen. The other three specimens (T1, T2, T3) contained copper shims before welding so that the shape of the copper after welding could be used to identify the action of the tool upon the material at the weld line (e.g. Fig. 3). The measured torque plotted against the axial distance moved by the tool from each of the test welds is presented in Fig. 4. The results show a spike of torque at first touchdown. During welding, the toque is around 40 ± 3 Nm at the start of welding and gradually drops to 37 ± 2 Nm towards the end of the weld length.

Model Description

The CFD model is quite straightforward and is shown in Fig. 5. Three regions are created representing the tool, the workpiece, and the backing plate.

The travel of the tool is captured by "moving" the workpiece. The volume of fluid method is used to separate the two sides of the workpiece, and to measure the mixing of the workpiece sides following welding. The tool is rotated at the given rotation rate, and the interface between the tool and the workpiece experiences the shear stress associated with that.

All simulations are steady state, which implies an infinite weld. Subsequently, the far-field temperature boundary conditions, which are fixed based on the test conditions, are adjusted to account for this compared to short test section.

The workpiece viscosity is included as outlined in the previous sections.

As discussed, a material behaviour model (Eqs. 1 to 4) has been developed to investigate the possibility of generating a material model for FSW that can be easily used for a wide range of materials without the need to either find esoteric material data or to perform a substantial series of material tests before a simulation can be performed. The only parameters required are the room temperature pseudo-static material strength (σ_{0ST}^{RT}) and the melting temperature (T_M) as demonstrated in Eq. 1. Assumed values were taken from a webpage of material data [7]. It was assumed

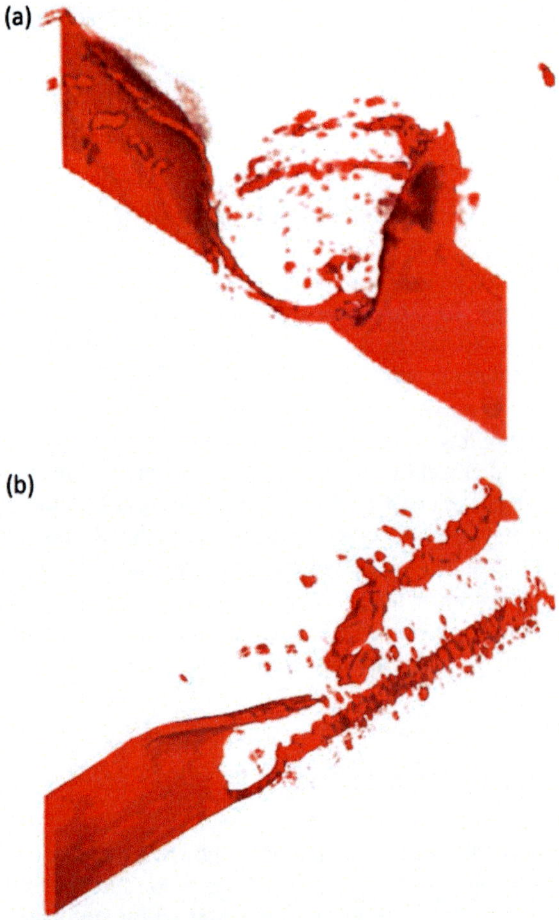

Fig. 3 X-ray CT images of the copper shims in two of the test specimens after welding. Specimen T3 in which the copper shim was situated transverse to the weld line (**a**) and specimen T1 in which the copper shim was situated along the axis of the weld line (**b**). (Color figure online)

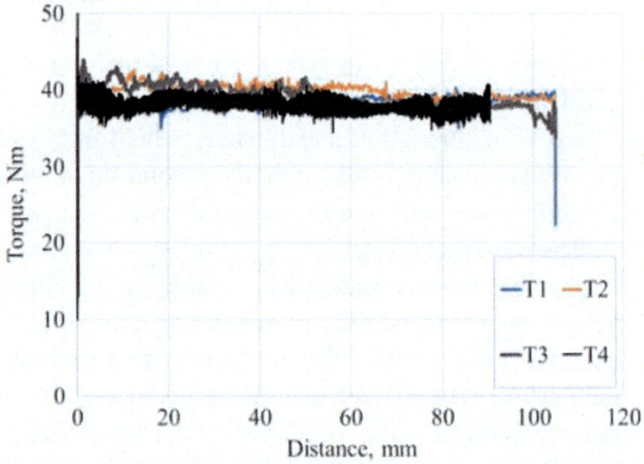

Fig. 4 Measured torque from all four tests plotted against tool travel. (Color figure online)

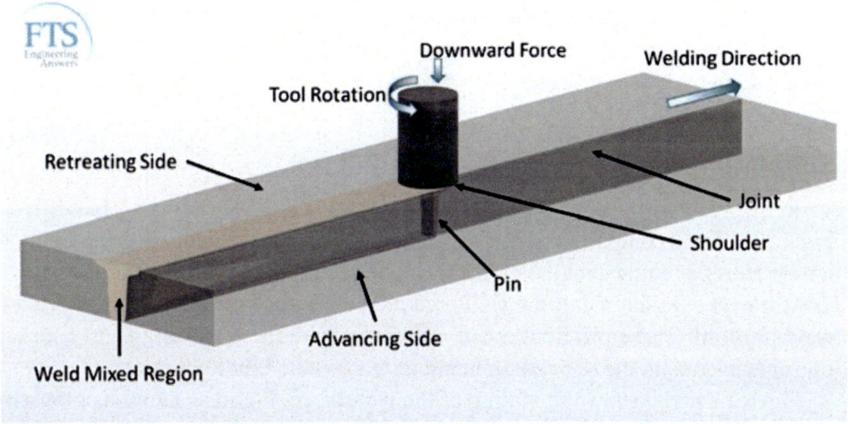

Fig. 5 Schematic of the CFD model. (Color figure online)

Fig. 6 Components of the assumed material strength curve based on the room temperature strength and the melting point of the material. (Color figure online)

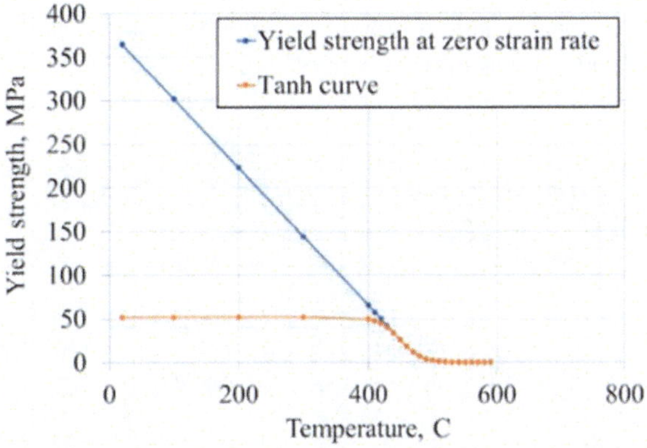

that the correct room temperature strength parameter to be used was the ultimate tensile strength which was 483 MPa. The melting temperature assumed was 502 °C. These values were adopted in Eq. 1 and the transition temperature (T_T) of Eq. 2 was assumed to be 450 °C. With these figures, strength curve assumed by the model is given in Fig. 6 (a linear curve below 450 °C and a tanh curve above it). This curve was processed using Eqs. 3 and 4 to give a viscosity for the analysis. This curve can easily be generated for any material and so the viscosity input for CFD models of FSW is straightforward to obtain.

A further and important aspect of the modelling undertaken is that the heat generation was assumed to be by adiabatic heating only. Many models of FSW adopt advanced friction characteristics (e.g. [8]) that would be both expensive and laborious to develop on a bespoke basis for general analyses of FSW. The current work assumed that friction was caused by the shearing of the workpiece material in the small crevices that represent the surface of the tool, so, even at a microscopic level friction is really the adiabatic shearing of material.

Simulation Results

Figure 7 shows a comparison of the test temperature measurements with the simulated temperatures at equivalent locations. It can be seen for the three locations compared that the peak measurements are well predicted with small differences away from the weld position. These differences are associated with the test being short timescale transient event and the simulation being a steady-state infinite weld.

Figures 8 and 9 show the mixing of the workpiece. Figure 8 shows a comparison between the simulation and test imaging of the copper shim used in the testing. It can be seen that the shim downstream of the tool is split into a top and bottom half which is also seen in the simulation. Figure 9 shows a contour plot of the mixing of the simulation. The retreating side of the workpiece is coloured dark grey, and the advancing side is coloured white. The mixing is shown by the colour range of reds and yellows downstream of tool. This can be used to assess the effectiveness of the weld, where improved welding can be seen by increased mixing.

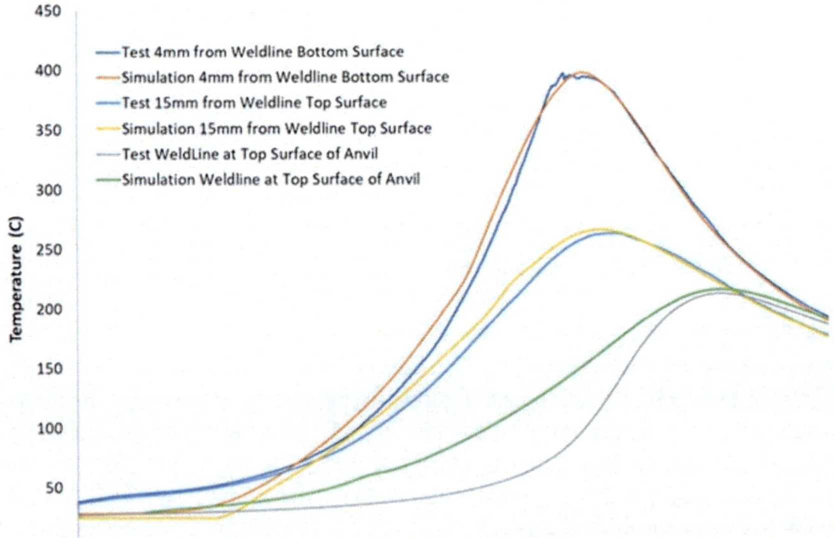

Fig. 7 Comparison of temperature measurements with simulated temperatures at three locations. (Color figure online)

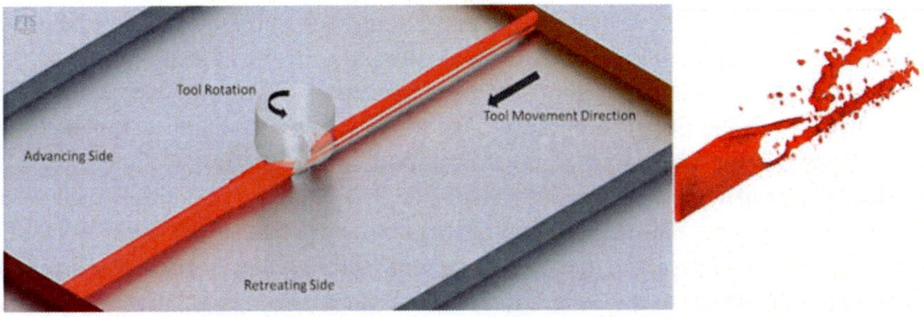

Fig. 8 Mixing of the tool piece: comparison of the simulation (left) with the test copper shim imaging (right). (Color figure online)

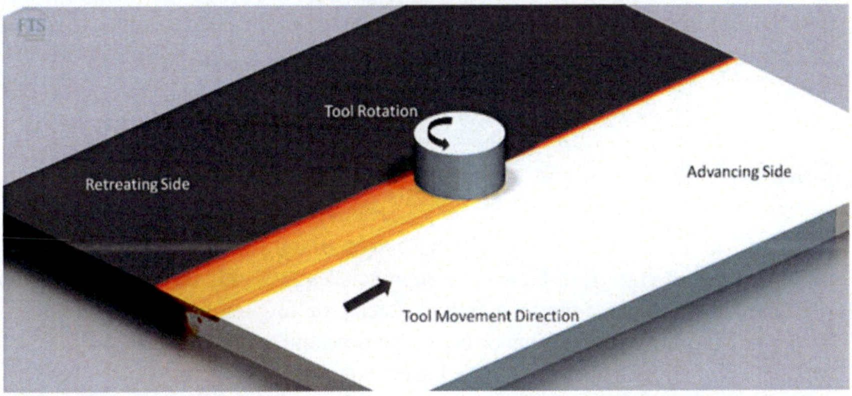

Fig. 9 Mixing of the tool piece: the retreating side of the workpiece is coloured dark grey, and the advancing side is coloured white. The mixing is shown by the colour range of reds and yellows downstream of tool. (Color figure online)

The torque was underpredicted by the simulation compared to test (30 Nm v 37 Nm) and subsequently the heat input was underpredicted (1800 W v 2300 W). These differences are anticipated to be because the steady-state simulation is not an exact representation of the test; as discussed earlier, the torque reduces during the test, and a longer weld would see a torque more in line with the simulation.

Discussion

The ability of mechanical and process engineers to adopt computer-based systems to answer the regular problems that they must solve has grown dramatically in only a few decades. Weld process modelling has been a more difficult aspect of engineering to analyse using computer models because the related physical processes are so complex. FSW is more amenable to analysis because the global material response is at temperatures that are below the melting point. This means that an analyst can assume slow viscous flow. The current paper addresses a possible barrier to the more widespread adoption of FSW in industry as part of a portfolio of techniques that could be used to answer questions about the possible use of FSW in any application. The widespread availability of both commercial software and consultants who can provide CFD facilities means that access to such a modelling capability is straightforward. It has been assumed that access to a substantial quantity of materials data would present a commercial resource limitation to the uptake of CFD by FSW engineers. The current study has attempted to provide a solution through the adoption of a simplified procedure that leads to the generation of the input data for an accurate CFD model of FSW based on only the room temperature, pseudo-static material strength, and its melting temperature. The validation study has shown that this technique is sufficiently accurate and, together with the availability of easy-to-use CFD software, it forms the basis of method that can be adopted by FSW process engineers to considerably develop and fine-tune their welding processes without the need to adopt what could be time-consuming in a costly welding trial.

Conclusions

A material model has been developed that can be straightforwardly used for as part of a simulation-based study of FSW. The approach requires only the room temperature, pseudo-static material strength, and the material melting temperature. The study has shown that the simple model can be used to understand the flow of material during the production of a FSW, and, as such, represents the basis of placing process modelling in the hands of FSW process engineers quickly and at low cost.

References

1. Mukherjee T, Wei HL, De A, DebRoy T (2018a) Heat and fluid flow in additive manufacturing—Part I: Modeling of powder bed fusion. Comput Mater Sci 150:369–313
2. Mukherjee T, Wei HL, De A, DebRoy T (2018b) Heat and fluid flow in additive manufacturing—Part II: powder bed fusion of stainless steel, and titanium, nickel and aluminum base alloys. Comput Mater Sci 150:369–380
3. Colegrove PA (2003) Modelling of friction stir welding, PhD thesis, The University of Cambridge. https://doi.org/10.17863/CAM.14008
4. A Askari, S Silling, B London, M Mahoney (2001) Modeling and analysis of friction stir welding processes, The Minerals, Metals & Materials Society
5. SIEMENS: Engineer innovation with CFD-focused multiphysics simulation. https://www.plm.automation.siemens.com/global/en/products/simcenter/STAR-CCM.html
6. Dickerson, T (first published 2002) The FSW benchmarks, Cambridge University Engineering Department. https://www-materials.eng.cam.ac.uk/FSW_Benchmark.
7. ASM Aerospace Specification Metals Inc (accessed April 2020): 'Aluminum 2024-T3'. http://asm.matweb.com/search/SpecificMaterial.asp?bassnum=MA2024T3
8. N Dialami, M Cervera, M Chiumenti, C Agelet de Saracibar (2017) A fast and accurate two-stage strategy to evaluate the effect of the pin tool profile on metal flow, torque and forces in friction stir welding. Int J Mech Sci 122:215–227

Effect of Tool Geometries on "Heat-Input" During Friction Stir Welding of Aluminum Alloys

Yutaka S. Sato, Yuichiro Tanai, Dalong Yi, Tianbo Zhao, and Hiroyuki Kokawa

Abstract Many previous studies proposed equations and models to estimate the heat-input of FSW, but their verifications would be still insufficient. In this study, the heat-input was calorimetrically measured during FSW of aluminum alloys with various welding tools having different shapes, and the effect of tool geometries on heat-input was examined. The measured heat-input increased with increasing shoulder diameter, which was in good agreement with many previous works. Interestingly, the heat-input strongly depended on the probe diameter and length. An attempt to correlate the heat-input with the tool surface area was made, virtually showing that the heat-input linearly increased with increasing "effective surface area" defined as (half of shoulder surface area + probe column surface area + probe tip surface area). This result experimentally suggested that the effect of probe dimension on heat-input was unexpectedly high. Moreover, an empirical equation estimating the heat-input during FSW of aluminum alloys from the welding parameters and the tool geometry was successfully established.

Keywords Friction stir welding · Aluminum alloys · Heat-input · Tool shape · Calorimetry

Introduction

There have been extensive studies on the fundamental nature of friction stir welding (FSW), such as microstructure evolution, material flow, heat generation and transfer, weld properties, and residual stresses. Heat generation (or heat-input) is believed to be one of the key factors governing the defect formation and their mechanical properties, and is well known to be very sensitive to the welding parameters. Many

Y. S. Sato (✉) · Y. Tanai · D. Yi · T. Zhao
Department of Materials Processing, Graduate School of Engineering, Tohoku University, 6-6-02 Aramaki-aza-Aoba, Aoba-ku, Sendai 980-8579, Japan
e-mail: ytksato@material.tohoku.ac.jp

H. Kokawa
Shanghai Key Laboratory of Materials Laser Processing and Modification, School of Materials Science and Engineering, Shanghai Jiao Tong University, Shanghai 200240, China

Y. Hovanski et al. (eds.), *Friction Stir Welding and Processing XI*, The Minerals, Metals & Materials Series, https://doi.org/10.1007/978-3-030-65265-4_16

attempts [1–5] have been made to achieve the accurate estimation of heat-input during FSW through the calculation of various methods, e.g., friction heat-based model [6], pseudo heat index method [7], torque-based model [8], and process force-based model [9]. However, FSW is a complicated process influenced by multiple factors, so that approaches to estimate the heat-input are still very difficult through calculation. Moreover, the estimated heat-inputs have not systematically been compared with the properties or the resulting microstructure because they were validated using the welds produced with the limited welding parameters.

To overcome the difficulty in the accurate estimation of heat-input during FSW, Aota et al. [10] proposed the calorimetric technique for the heat-input measurement during FSW of aluminum alloy. FSW was conducted in the specially designed water-filled box, and the generated heat-input was directly measured from the change in the water temperature. They successfully estimated the heat-inputs during FSW and gas metal arc (GMA) welding, and showed that the heat-input during FSW was approximately half of that of GMA welding. However, they did not establish any equations of the heat-input with the welding parameters because of the limited number of measurements. Using the calorimetric technique proposed by Aota et al., the authors also measured the heat-input experimentally during FSW of aluminum alloys 1100, 5083, and 6063 at the wide range of welding parameters in the previous studies, and established the empirical equation estimating the heat-input with the welding parameters [11–13].

Besides the welding parameters, the tool geometry is also a significant factor to control the heat generated during FSW. Many previous studies [4, 14–16] have mainly focused on the contribution of the shoulder features to the heat generation because it is believed that the shoulder generated more than 80% of total heat generation. Recently, numerical FSW models have been developed to examine the heat generation from the probe during FSW. Using a commercial finite element package, Al Bhadle et al. [17] showed the amount of heat generation increased the number of flats on the tapered tool profile. Leon et al. [18] developed an analytical heat-input model, reporting that the total heat supply increased with increasing probe diameter, although the tool probe has limited contribution to the heat generation, compared to the shoulder. These results are useful to understand contribution of tool geometry to heat generation during FSW, but the effect of tool geometries on heat-input has hardly been examined systematically.

In this study, the heat-inputs during FSW of aluminum alloys were experimentally measured with various welding tools with different shapes. Effect of tool geometries on measured heat-input was examined, and an empirical equation estimating the heat-input with the welding parameters and a tool geometric factor attempted to be established using all experimental data obtained during FSW of three aluminum alloys.

Experimental Procedures

The base materials used in this study were commercial aluminum alloys 1100-O, 5083-O, and 6063-T5, 4 mm in thickness. A schematic illustration of the apparatus used for the heat-input measurements in this study is shown in Fig. 1a. The backing plate made of copper and the workpiece were placed in a water-filled acrylic box. The bead-on-plate FSW was carried out in all the cases. To examine the effect of tool geometries on heat-input during FSW of aluminum alloys, the six welding tools having cylindrical probes with threads were made of the tool steel. Dimensions of the welding tools used in this study are listed in Table 1. The appearance of the tool T1 is shown in Fig. 1b. The welding tool was tilted by 3° from the workpiece normal during FSW. The tool rotational speed varied from 1000 to 3000 rpm, and the travel speed ranged between 200 mm/min and 600 mm/min.

During the calorimetric measurements, water was poured into the acrylic box prior to FSW and kept at a level with the backing plate to prevent it from boiling when in contact with the workpiece. In order to reach a temperature equilibrium quickly, the water was stirred during the experiment. The water temperature was continuously

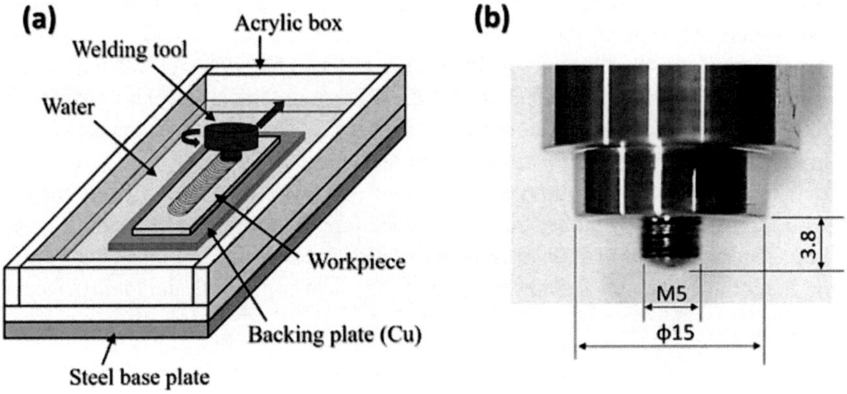

Fig. 1 **a** Schematic illustration of apparatus used for measurement of heat-input during FSW and **b** appearance of the welding tool used

Table 1 Dimensions of the welding tools used in this study

Tool ID	Shoulder diameter (mm)	Probe diameter (mm)	Probe length (mm)
T1	15	5	3.8
T2	15	8	3.8
T3	17.5	5	3.8
T4	20	5	3.8
T5	20	8	3.8
T6	20	8	5.8

recorded using six K-type thermocouples. The heat-input was calculated using the following equation:

$$\mathrm{HI} = \left[\left(c_{\mathrm{wa}} m_{\mathrm{wa}} + c_{\mathrm{bp}} m_{\mathrm{bp}} + c_{\mathrm{wp}} m_{\mathrm{wp}}\right) \cdot \Delta T\right] / L \quad (1)$$

where HI is the welding heat-input (J/mm); c_{wa}, c_{bp}, and c_{wp} are the specific heat of water, backing plate, and workpiece (J/K kg), respectively; m_{wa}, m_{bp}, and m_{wp} are the weights of water, backing plate, and workpiece (kg), respectively; ΔT is the temperature difference of the water, backing plate and workpiece (K) before and after welding; and L is the welding distance (mm).

Results and Discussion

The measured heat-input increased with increasing rotational speed and with decreasing travel speed, which was in good agreement with many previous studies. The effect of alloy type on heat-input is shown in Fig. 2. Aluminum alloys 1100 and 6063 exhibited much higher heat-input than aluminum alloy 5083 at the same welding parameters. The reason for this result is still unclear, but this could be explained by the difference in the thermal conductivities of these alloys (AA1100: 222 W/m K, AA5083: 117 W/m K, AA6063: 209 W/m K at room temperature) because the heat might generate more so as to compensate for the heat dissipated into the base material in the alloys with the higher thermal conductivities.

The heat-input was remarkably affected by the tool shape, i.e., it increased with increasing the shoulder diameter, probe length, and probe diameter. Since the heat generation during FSW is associated with friction and/or deformation heating on the tool surface, an attempt to correlate the heat-input with the tool surface area

Fig. 2 Effect of alloy type on measured heat-input

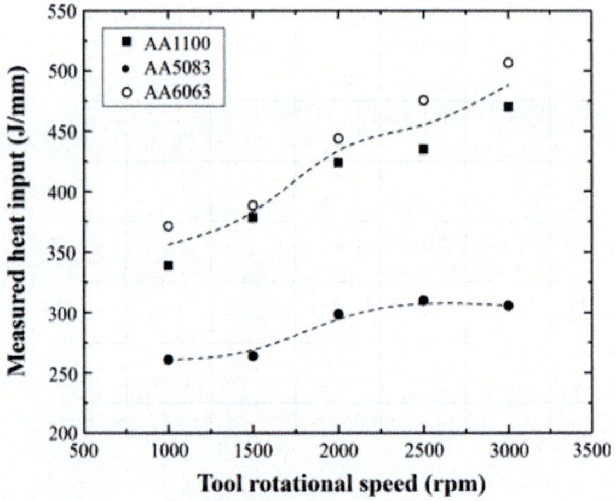

(= shoulder surface area + probe column surface area + probe tip surface area) was made. Effect of tool surface area on measured heat-input during FSW of aluminum alloy 5083 is shown in Fig. 3. It is found that the heat-input increases with increasing tool surface area, but a simple correlation between the heat-input and the tool surface area could not be obtained. Interestingly, an increase in probe diameter (*d*) increases the heat-input more than an increase in shoulder diameter (*D*), although an increase in tool surface area with increasing probe diameter (*d*) is much smaller than that with increasing shoulder diameter (*D*). This result implies that an increase in tool surface area with increasing shoulder diameter (*D*) is too large for an increase in heat-input.

To reduce the increase in tool surface area with increasing shoulder diameter, this study defined "effective surface area" as (half of shoulder surface area + probe column surface area + probe tip surface area). Effect of effective surface area on heat-input during FSW of aluminum alloy 5083 is presented in Fig. 4. The heat-input linearly increases with increasing effective surface area. This result suggests that the probe dimension contributes to the heat-input strongly, possibly more than

Fig. 3 Effect of total surface area on measured heat-input during FSW of AA5083. In this figure, *D* is shoulder diameter, *d* is probe diameter, and *h* is probe height

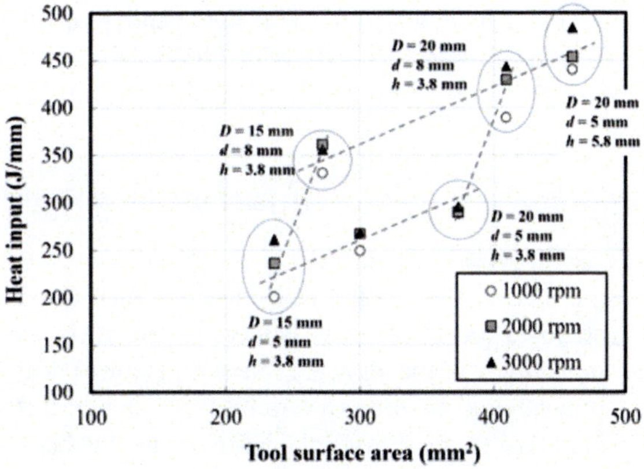

Fig. 4 Effect of effective surface area on measured heat-input during FSW of AA5083

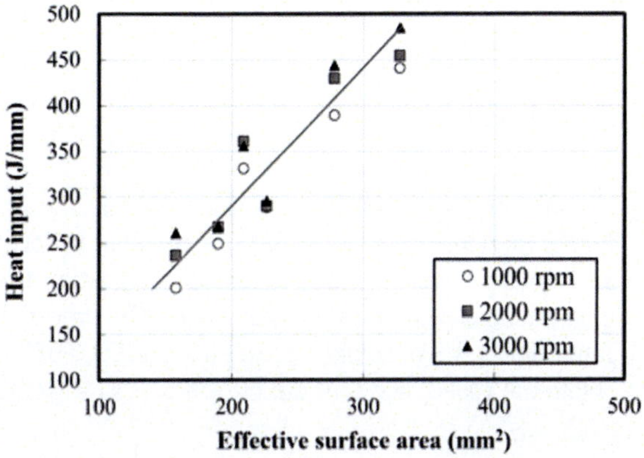

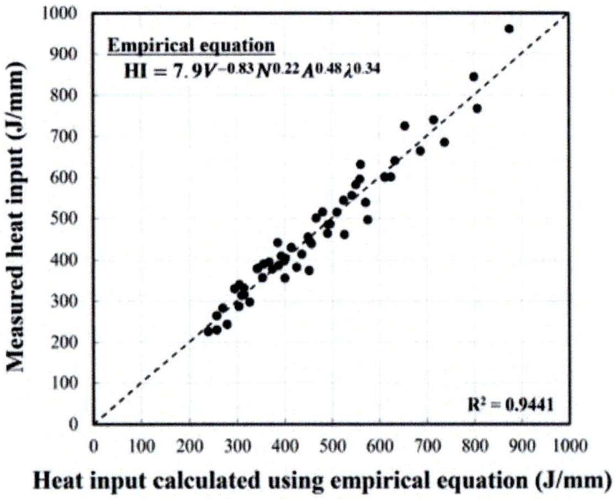

Fig. 5 Relationship between measured heat-input and heat-input calculated using the empirical equation

the shoulder, although many previous papers [4, 14–17] mentioned that the shoulder was the main contributor to the heat generation during FSW.

Using all data obtained during FSW of aluminum alloys 1100, 5083, and 6063, the multiple regression analysis established the following empirical equation estimating the heat-input with the welding parameters and the tool dimension:

$$\mathrm{HI} = 7.9V^{-0.83}N^{0.22}A^{0.48}\lambda^{0.34} \quad (2)$$

where HI is the welding heat-input (J/mm), V is the travel speed (mm/s), N is the rotational speed (rad/s), A is the effective surface area (mm^2), and λ is the thermal conductivity of the alloy at room temperature (W/m K). Based on Fig. 2, the thermal conductivity λ is used as a variable reflecting the difference in alloy type. The relationship between the measured heat-input and the heat-input calculated using the empirical equation is presented in Fig. 5. They exhibit a good correlation, i.e., the empirical equation can be used to estimate the heat-input during FSW of three aluminum alloys.

Conclusions

The higher heat-input was measured during the FSW of aluminum alloy with the higher thermal conductivity. A good correlation between the heat-input and "effective surface area" was virtually obtained, suggesting that the effect of probe dimension on heat-input was unexpectedly high. The empirical equation estimating the heat-input during FSW of aluminum alloys 1100, 5083, and 6063 from the travel speed, rotational speed, and tool dimension was successfully established.

Acknowledgements The authors are grateful to Mr. A. Honda, Ms. F. Yu, Mr. W.Y. Park and Mr. P. Predeep for their technical assistance. They also wish to thank Dr. S. Mironov, Prof. A. Kawasaki, Prof. K. Anzai and Prof. K. Yoshimi for their helpful discussion. This work was partially supported by JSPS KAKENHI (a Grant-in-Aid for Challenging Exploratory Research, Grant No. 15K14137). This work was partially performed under the Joint Collaboration Program of Joining and Welding Research Institute, Osaka University.

References

1. Balasubramanian N, Gattu B, Mishra RS (2009) Process forces during friction stir welding of luminium alloys. Sci Technol Weld Join 14:141–145
2. Reynolds AP (2007) Microstructure XE "microstructure" development in aluminum alloy friction stir welds. In: Mishra RS, Mahoney MW (eds) Friction stir welding and processing. ASM International, Materials Park, Ohio, pp 51–70
3. Khandkar MZH, Khan JA, Reynolds AP (2003) Prediction of temperature distribution and thermal history during friction stir welding: input torque based model. Sci Technol Weld Join 8:165–174
4. Schmidt H, Hattel J, Wert J (2004) An analytical model for the heat generation in friction stir welding. Modell Simul Mater Sci Eng 12:143–157
5. Colegrove PA, Shercliff HR (2003) Experimental and numerical analysis of aluminium alloy 7075–T7351 friction sstir welds. Sci Technol Weld Join 8:360–368
6. Frigaard Ø, Grong Ø, Midling OT (2001) A process model for friction stir welding of age hardening aluminum alloys. Metall Mater Trans A 32A:1189–1200
7. Arbegast WJ (2003) Modeling friction stir joining as a metal working process. In: Jin Z et al (eds) Hot deformation of aluminum alloy III. TMS, Warrendale, PA, pp 313–327
8. Pew JW, Nelson TW, Sorensen CD (2007) Torque based weld power model for friction stir welding. Sci Technol Weld Join 12:341–347
9. Kumar R, Singh K, Pandey S (2012) Process forces and heat input as function of process parameters in AA5083 friction stir welds. Trans Nonferrous Met Soc China 22:288–298
10. Aota K, Okamura H, Ezumi M, Takai H (2001) Heat inputs and mechanical properties in friction stir welding. In: Proceedings of 3rd International FSW Symposium, Kobe, Japan, CD-ROM
11. Sato YS, Onuma T, Ikeda K, Kokawa H (2016) Experimental verification of heat input during friction stir welding of Al alloy 5083. Sci Technol Weld Join 21:325–330
12. Yi D, Onuma T, Mironov S, Sato YS, Kokawa H (2017) Evaluation of heat input during friction stir welding of aluminium alloys. Sci Technol Weld Join 22:41–46
13. Zhao T, Sato YS, Kokawa H, Ito K (2020) Predicting tensile properties of friction-stir-welded 6063 aluminum with experimentally measured welding heat input. Acta Metall Sinica 33:1235–1242
14. Tang W, Guo X, McClure JC, Murr LE, Nunes A (1998) Heat input and temperature distribution in friction stir welding. J Mater Process Manuf Sci 7:163–172
15. Neto DM, Neto P (2013) Numerical modeling of friction stir welding process: a literature review. Int J Adv Manuf Technol 65(1–4):115–126
16. Schmidt HB, Hattel JH (2008) Thermal modelling of friction stir welding. Scripta Mater 58:332–337
17. Al Bhadle BMA, Al Azzawi RAA, Thornton R, Beamish K, Shi S, Dong HB (2019) Equations of heat generation during friction stir welding for tapered polygonal tools. Sci Technol Weld Join 24:93–100
18. Leon SJ, Jayakumar V (2020) Transient heat input model for friction stir welding using non-circular tool pin. FME Trans 48:137–142

Experimental and Numerical Investigations of High Strain Rate Torsion Tests of Al-Based Alloys at Elevated Temperatures

Anton Naumov, Anatolii Borisov, and Anastasiya Borisova

Abstract This study aims to validate a finite element model (FEM) of the high strain rate torsion tests on several aluminum-based alloys by realizing physical simulation of FSW by torsion tests, finite element modelling of this process, and comparing the obtained data. A high strain rate torsion test at elevated temperatures was realized on Gleeble-3800 system for different Al alloys: AA 1050, AA 5082 O, and AA 2024 T4. High strain rate torsion tests were performed at two elevated temperature—350 and 500 °C. The stress–strain curves were obtained and analyzed after physical simulation. A finite element model was built and validated using temperature and torque data obtained after physical simulation. The stress–strain curves obtained by numerical simulation are in good agreement with the experimental data.

Keywords Al-based alloys · Friction stir welding · Finite element modelling · Physical simulation · High strain rate · Torsion test

Introduction

Friction stir welding (FSW) is a prospective and relatively new method for joining materials in solid state [1]. This technique is widely used to produce welded structures from aluminum alloys and provides better joint properties compared to fusion methods. A unique aspect of joint properties in FSW is due to the grain size refinement as a result of severe plastic deformation at elevated temperatures [2–7].

The study of the stress–strain characteristics of the FSW process is an urgent task for the researchers in the fields of material science and mechanical engineering. These characteristics, on the one hand, have a great influence on heat generation, and on the other hand on the weld microstructure and mechanical properties evolution. However, it is difficult to evaluate these parameters using FSW process itself; therefore, researchers pay special attention to the physical and numerical simulation

A. Naumov (✉) · A. Borisov · A. Borisova
Peter the Great St. Petersburg Polytechnic University, St. Petersburg, Russia
e-mail: anton.naumov@spbstu.ru

Y. Hovanski et al. (eds.), *Friction Stir Welding and Processing XI*, The Minerals, Metals & Materials Series, https://doi.org/10.1007/978-3-030-65265-4_17

of FSW to estimate or calculate stress–strain parameters [8–12]. FSW is characterized by both a high amount and a high rate of strain at elevated temperatures, which are difficult to implement even on the most modern systems for physical simulation (PS). From the experience of the researchers involved in the physical simulation of the FSW, it can be assumed that it is possible to evaluate the parameters of FSW using torsion experiments [13–15]. In the present research, the high strain rate torsion tests at elevated temperatures were completed to physically simulate the deformation processes of the stir zone during FSW.

Numerical simulation (NS) is an important part of product development cycle today. This study aims to validate the finite element model (FEM) of the high strain rate torsion tests on several aluminum-based alloys by realizing the physical simulation of FSW by torsion tests, finite element modelling of this process, and comparing the obtained data.

Experimental Procedure

Materials

The materials used for the physical and numerical simulation in the present research include the following Al alloys: AA 1050, AA 5082 O, and AA 2024 T4, which are described in detail in the previous papers [16, 17]. The Al alloys' chemical composition is shown in Table 1.

Physical Simulation

Physical simulation of high strain rate torsion was performed on a Gleeble-3800 system and a Torsion mobile convert unit. The standard torsion specimen has a working diameter of 10 mm. To achieve higher cooling rates after hot deformation, the diameter of the sample working space was reduced to 5.7 mm. A dimensioned drawing of sample for the physical simulation is shown in Fig. 1. A thermocouple was welded to the fixed part of the sample to control and monitor the temperature during torsion.

Table 1 Chemical compositions of researched aluminum alloys, mass. %

	Si	Fe	Cu	Mn	Mg	Cr	Zn	Ti
1050	0.15	0.20	–	0.03	–	–	0.05	–
5082	0.24	0.25	0.06	0.44	4.8	0.03	0.09	0.04
2024	0.19	0.13	4.45	0.36	1.24	–	–	0.03

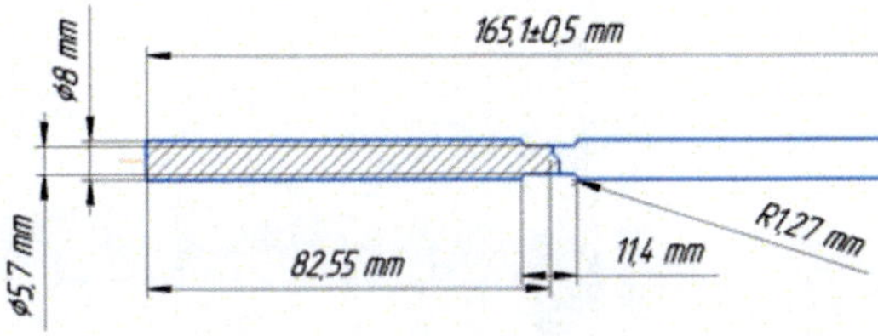

Fig. 1 The sample for physical simulation. (Color figure online)

During the experiments, the moment and rotation of the traverse were recorded. Using these data, the equivalent stress and strain were calculated using the following equations:

$$\overline{\varepsilon} = \frac{1}{\sqrt{3}} \cdot \frac{r \cdot \theta}{l}, \tag{1}$$

where θ is the number of rotations in radians, r and l are the radius and length of the working space in mm;

$$\overline{\sigma} = \sqrt{3} \cdot \frac{M}{2 \cdot \pi \cdot r^3} \cdot (3 + n + m), \tag{2}$$

where M is the moment in N·m, n is the coefficient of strain hardening, m is the coefficient of strain rate, $n = m = 0$.

Numerical Simulation

The basic principles of numerical simulation in Deform™ software are described in previous articles [16, 17]. The standard base grades from the software database with material properties were used for the simulation. Here are some features of the model:

- the material of the sample in the initial state (before deformation) is isotropic and there are no initial stresses and deformations;
- the material of the sample in the initial state is accepted as plastic;
- for the simplicity of the model during numerical simulation, it is presumed that processes outside the working space may be neglected and the simulation starts after the heating but before the torsion.

The scheme of the sample model is shown in Fig. 2.

Physical and numerical simulation was carried out at various rotation speeds, temperatures, and numbers of revolutions. The experimental matrix for each alloy is presented in Table 2, and the number of rotations is indicated in the columns.

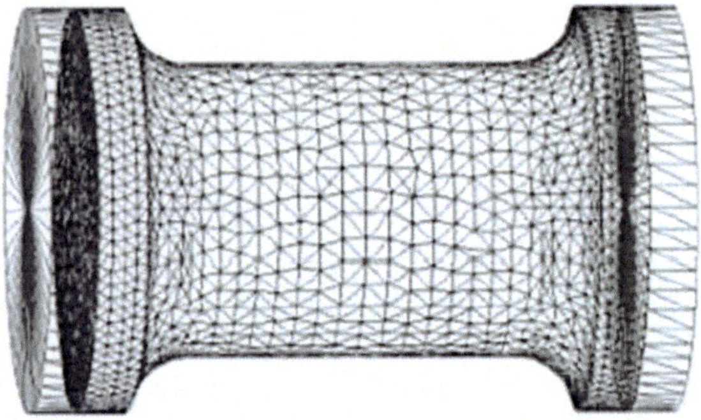

Fig. 2 Numerical model from left to right: fixed disc, meshed working zone, moving disk

Table 2 The experimental matrix

T\Rotation speed (°C)	700 rpm	1000 rpm
500	1 and 1.5	1 and 1.5
350	1 and 1.5	1 and 1.5

Results and Discussion

According to the results of physical and numerical simulation, the dependences of the effective stress from the effective strain during torsion are obtained. The effective stress–effective strain curves for all alloys are presented in Fig. 3. The comparison between experimental and calculated results is shown in Figs. 4, 5 and 6.

As expected, alloy AA 1050 has the lowest stress level in comparison to other alloys while the alloy AA 5082 O has the highest stress level at a constant temperaure. It is shown that at temperature of 500 °C the stress level remains almost unchanged throughout the deformation, and it is less than the stress level at a temperature of

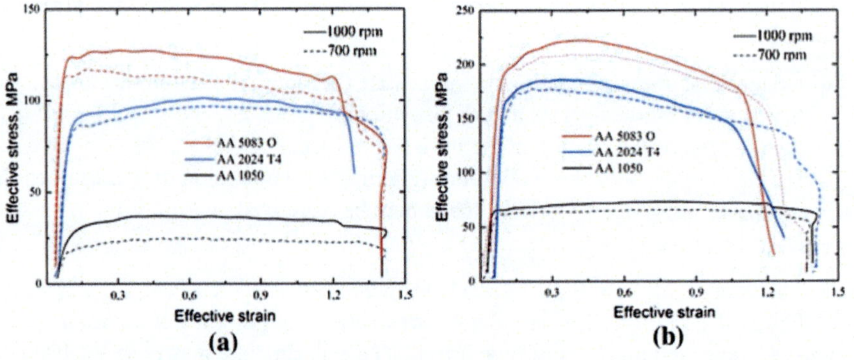

Fig. 3 Effective stress–effective strain curves during torsion of alloys AA 1050, AA 5082 O, and AA 2024 T4 at **a** 500 °C, **b** 350 °C. (Color figure online)

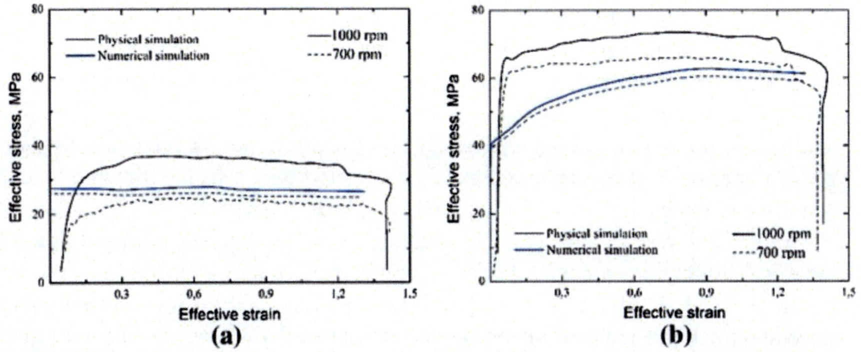

Fig. 4 Effective stress–effective strain curves during torsion of AA 1050 alloy at **a** 500 °C, **b** 350 °C. (Color figure online)

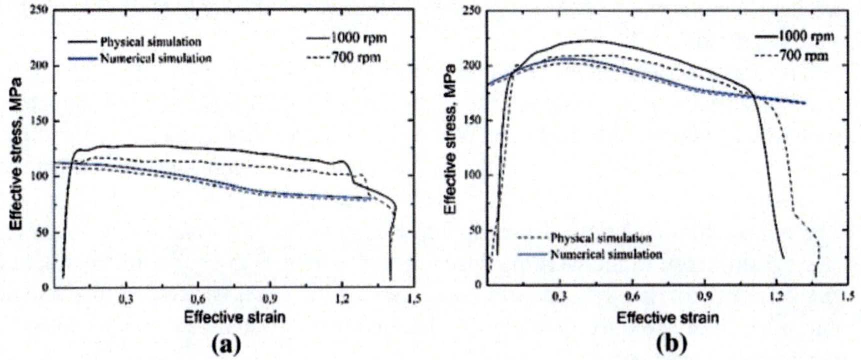

Fig. 5 Effective stress–effective strain curves during torsion of AA 5082 O alloy at **a** 500 °C, **b** 350 °C. (Color figure online)

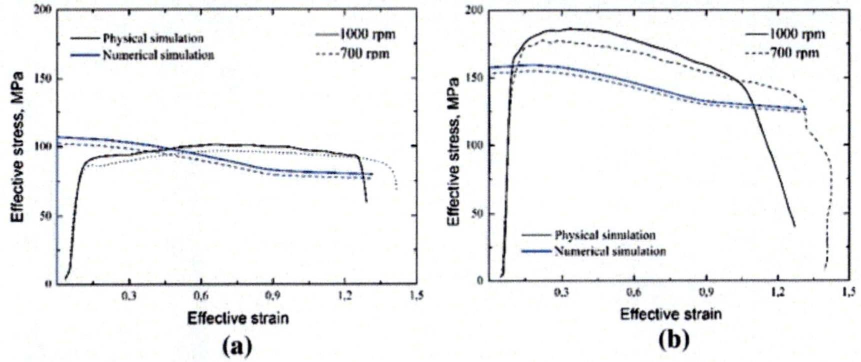

Fig. 6 Effective stress–effective strain curves during torsion of AA 2024 T4 alloy at **a** 500 °C, **b** 350 °C. (Color figure online)

350 °C for all alloys. While at T = 350 °C alloys AA 2024 T4 and AA 5082 O have a significant decrease in stress level after reaching a strain value of about 0.4. For AA 5082 O, the difference between the maximum stress level and minimum stress level is about 50 MPa, for alloy AA 2024 T4—about 40 MPa. Similar values were obtained during numerical simulation. Stress relaxation may be related to dynamic recrystallization or recovery processes. For all performed tests the maximum stress values were obtained at T = 350 °C, rotation speed—1000 rpm.

The obtained results show that the nature of the curves obtained after physical and numerical simulation coincides, but the obtained stress values are different. However, during physical simulation the stress values were recorded for the whole working space of the sample, meanwhile during numerical simulation, we are able to estimate the stress level for each node of the finite element mesh. Therefore, in this study, the point P2 in the center of the sample working space surface, shown in Fig. 7, was chosen to build the effective stress —effective strain curves. In the center of the working space surface, according to the simulation results, the highest stress values were observed.

The importance of physical simulation of FSW, in this torsion test, is the possibility of quenching the sample directly after torsion to "freeze" the microstructure state which is impossible during FSW even using stop-action method. Using torsion tests at elevated temperatures with certain strain and strain rate parameters, the microstructure evolution of different weld zones during FSW could be analyzed. The numerical simulation of the torsion tests gives the information on strain and stress distribution in the working space of the sample. Subsequent microstructural analysis after torsion tests enables correlation of the stress–strain distribution with microstructural evolution. Numerical simulation of torsion tests could reduce the number of experiments.

Received stress–strain curves after physical simulation could be used as input materials properties for numerical simulation of FSW process thus increasing the predictive capability of the numerical simulation.

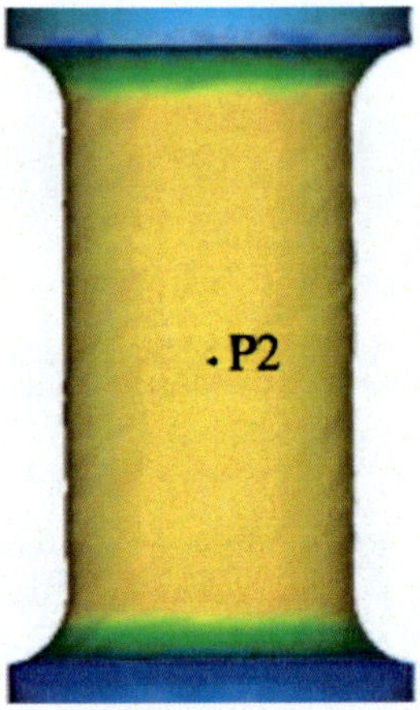

Fig. 7 Point P2 on the surface of the sample to obtain the level of stress values for material AA 1050, 350 °C, 700 rpm. (Color figure online)

Conclusions

The numerical simulation of the torsion tests gives the information on strain and stress distribution in the working space of the sample and also could reduce the number of experiments.

The results of the numerical simulation are in good agreement with the results of physical simulation. As was expected, alloy AA 1050 has the lowest stress level in comparison to other alloys, the alloy AA 5082 O has the highest stress level. For all performed tests the maximum stress values were obtained at T = 350 °C, rotation speed—1000 rpm.

Further rcsearch will be aimed to find the relationship between the microstructure and thermo-deformation parameters of the high strain rate torsion tests at elevated temperatures and the applicability of the received data to describe the thermo-deformation parameters that occur during FSW.

Acknowledgements The results of the work were obtained using computational resources of Peter the Great Saint-Petersburg Polytechnic University Supercomputing Center (www.spbstu.ru).

References

1. Dawes CJ, Thomas WM (1996) Friction stir process welds aluminum alloys. Weld J 75:41–45
2. Su L, Lu C, Li H, Deng G, Kiet Tieu A (2014) Investigation of ultrafine grained AA1050 fabricated by accumulative roll bonding. Mater Sci Eng A 614:148–155
3. Snopiński P, Król M (2018) Microstructure, mechanical properties and strengthening mechanism analysis in an AlMg5 aluminium alloy processed by ECAP and subsequent ageing. Metals 8:969
4. Mohamed IF, Masuda T, Lee S, Edalati K, Horita Z, Hirosawa S, Matsuda K, Terada D, Omar MZ (2017) Strengthening of A2024 alloy by high-pressure torsion and subsequent aging. Mater Sci Eng A 704:112–118
5. Rudskoy AI, Naumov AA, Chernikov EV (2014) Friction stir processing of metals is a new method of intensive plastic deformation. Tsvetnye Metally 4:36–40
6. Parshikov RA et al (2016) Analysis of specimen plastic flow features during severe plastic deformation. Rev Adv Mater Sci 45(1–2):67–75
7. Rudskoy AI, Zolotov AM, Ganin SV (2014) Modeling of process of equal channel angular pressing of billets, obtained from aluminum-based powders and rare-earth metals, in Capsules. Tsvetnye Metally 4:30–35
8. Masaki K, Sato Y, Maeda M, Kokawa H (2008) Experimental simulation of recrystallized microstructure in friction stir welded Al alloy using a plane-strain compression test. Scripta Mater 58:355–360
9. Livingston J, Lippold J (2013) Microstructure evolution during friction stir processing and hot torsion simulation of Ti-6Al-4V. Metall Mater Trans A 44:3815–3825
10. Golubev I, Morozova I, Naumov A, Hantelmann C, Doynov N, Michailov V (2017)Numerical simulation and experimental investigation on impulse friction stir welding of 6082-T6 aluminum alloy. Mater Sci Technol Conf Exhibit MS and T, 2:987–994
11. Chernikov E, Naumov A, Golubev I, Michailov V (2015)Numerical simulation and microstructure analysis of friction stir welded Al 6082. In: METAL 2015—24th International conference on metallurgy and materials, conference proceedings, pp 1423–1428

12. Zhang S et al (2018) Numerical analysis and analytical modeling of the spatial distribution of heat flux during friction stir welding. J Manuf Proces 33:245–255
13. da Fonseca EB, Santos TFA, Button ST, Ramirez AJ (2016) Physical simulation of a duplex stainless steel friction stir welding by the numerical and experimental analysis of hot torsion tests. Metallurg Mater Trans 47:4543–4552
14. Rule J, Lippold J (2013) Physical simulation of friction stir welding and processing of nickel-base alloys using hot torsion. Metall Mater Trans A 44:3649–3663
15. Poletti C, Krumphals F, Mitsche S, Gao Z (2013) Microstructural evolution of AA6082 with small aluminides under hot torsion and friction stir processing. Mater Sci Forum, T 753:263–266
16. Naumov A, Borisov A, Doroshchenkova A (2019) Numerical simulation of hot high strain rate torsion tests for al-based alloys. Key Eng Mater 822:66–71
17. Naumov A, Wenzel H, Borisov A, Doroshchenkova A (2019) Physical and numerical simulation of friction stir welding for Al-based alloys, MS and T 2019 - Materials Science and Technology 2019, pp 996–1001

Part VII
Spot Technologies

Advances in Refill Spot Welding Productivity

Yuri Hovanski, Andrew Curtis, Sarah Michaelis, Paul Blackhurst, and Brigham Larsen

Abstract With more than a decade of research and development documenting both the benefits and the challenges of refill friction stir spot welding (RFSSW), we see remarkedly few industrial applications that have moved to full implementation. Two major challenges have been recognized that are currently impeding further adoption; namely, cycle time and tool life. While these two challenges are not mutually exclusive, the focus of the current work will be to contest reported trends that claim reducing cycle time below 2 s has a deleterious effect on weld properties. Recently published work challenged this trend for RFSSW of a work-hardenable AA5052-H36, where weld times were reported as low as 800 ms (Larsen and Hovanski in Reducing cycle times of refill friction stir spot welding in automotive aluminum alloys, SAE International, 2020). While this study demonstrated that machine capabilities drive reduction in weld cycle time without a commensurate reduction in weld properties, the focus was on demonstrating these changes for a work-hardenable alloy rather than a precipitation-hardened alloy. As such, the current work will demonstrate that a similar reduction in cycle time may be realized in AA7075-T6.

Keywords Refill friction stir spot welding · Aluminum alloys · AA7075-T6 · Cycle-time

Introduction

Numerous studies have evaluated the RFSSW process in AA7075-T6. One of the earliest investigations of the RFSSW in this alloy was made by Shen et al. [2] in which an investigation of parameters on the microstructure and failure mechanisms of RFSSW joints in AA7075-T6 was presented. Shen's team investigated weld cycle times ranging from 3 to 5 s in duration with a range of rotational speeds from 1500 to 2000 RPM. Their investigation concluded that reducing the rotational velocity of the tool in combination with reducing the overall duration of the weld may lead

---

Y. Hovanski (✉) · A. Curtis · S. Michaelis · P. Blackhurst · B. Larsen
Department of Manufacturing Engineering, Brigham Young University, Provo, UT 84602, USA
e-mail: yuri.hovanski@byu.edu

Y. Hovanski et al. (eds.), *Friction Stir Welding and Processing XI*, The Minerals, Metals & Materials Series, https://doi.org/10.1007/978-3-030-65265-4_18

to preferable mechanical properties; however, they also concluded that increased rotational velocities at a longer duration may also lead to improvements in properties. In later work, Shen reported a direct correlation between material thickness and an increase in welding time and plunge depth, specifically concluding that decreasing weld time and plunge depth led to problems with bonding [3].

In a more recent study, Kubit et al presented data on RFSSW of 0.8 mm thick AA7075-T6 [4, 5], evaluating weld cycle times ranging from 1.5 s to 3.5 s. Like Shen et al., Kubit's team concluded that an increase in rotational speed and cycle time led to an increase in mechanical properties. However, in their investigation of failure mechanisms, they also recognized that either too much rotational velocity or a significant increase in the weld cycle time also reduced the overall performance of the RFSSW joints due to overheating. Furthermore, their more detailed investigation of plunge depth allowed them to conclude that shallow plunge depths resulted in interfacial fracture, while more significant plunge depths led to a nugget pullout fracture.

Others have investigated the process on AA7075-T6 at fixed welding speeds in which the influence of rotational velocity, plunge depth, and material surface conditions was investigated [6, 7]. Andres's team demonstrated that too low of a rotational velocity led to insufficient mixing, while too high led to the formation of voids along the bottom of the tool. This study was completed at a fixed weld time, so no conclusion regarding cycle time was included in this study.

A recent investigation of more than 100 papers focused on RFSSW showing an overall trend of reducing lap-shear strengths with a reduction in cycle time. The data showed in Fig. 1 summarizes results from 35 papers that report both strength and cycle time of RFSSW in various aluminum alloys and thicknesses. While the data covers spot sizes from 7 to 9 mm and thicknesses ranging from 0.8 mm to 3.5 mm, the overall observation of RFSSW strengths vs. cycle time may suggest a downward trend functionally related to the overall duration or cycle time of the joining process. Our purpose is to elucidate whether this trend is an accurate representation of the process capabilities, or if the reported data doesn't fully represent the overall capabilities of RFSSW.

Since the RFSSW process is more complicated than traditional friction stir welding and friction stir spot welding, the idea of cycle time is somewhat more complicated as well. Some authors report that weld duration is including the entire time the welding process is forging the materials together in the solid-state [8]; however, many fail to report the overall cycle time or weld duration, reporting only the plunge portion of the process or failing to account for dwelling that may occur at various stages of the overall process. For the purposes of this paper, we define the cycle time as phases C–H of Fig. 2. While manufacturing process calculations would need to accommodate for phases A-I for a full cycle time analysis, for comparative purposes with other work, our focus is on the process cycle time, also referred to as weld duration.

A more detailed look at cycle time of RFSSW in AA7075-T6 is presented in Fig. 3, where sheet thickness of each weld is displayed next to each point. This graph, similar to Fig. 1, shows no data below 1.5 s cycle times, due to the observation of the authors

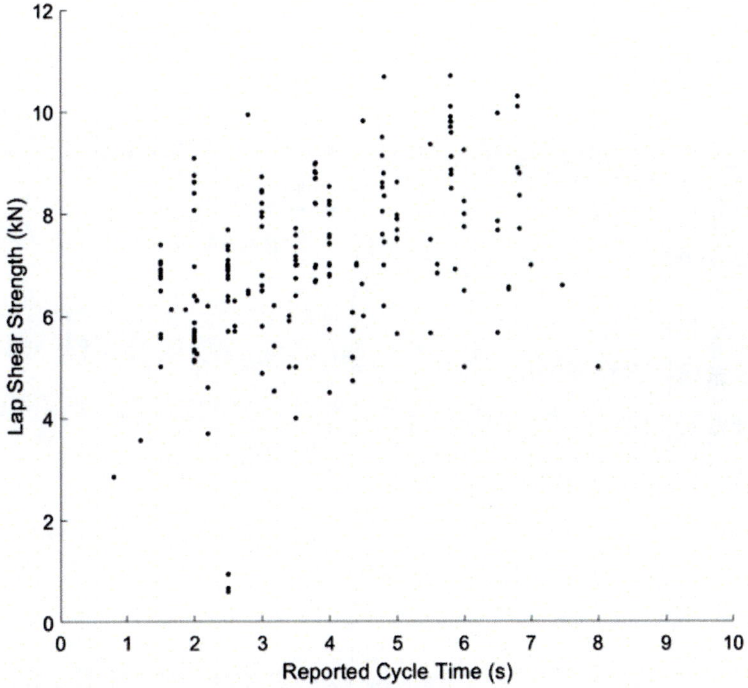

Fig. 1 Plot showing the relationship between RFSSW cycle time and lap-shear strength of RFSSW joints [2–6, 9–26]

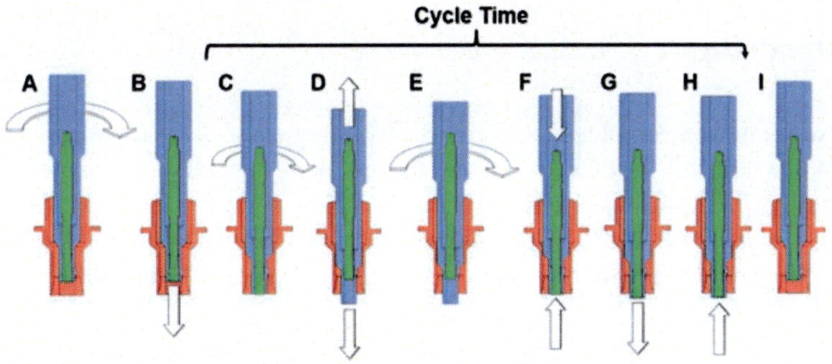

Fig. 2 A visual representation of the RFSSW process, **A** Start of rotation, **B** Clamping to tool touch-off, **C** Weld start, **D** Shoulder plunge/probe retract, **E** Dwell at depth, **F** Refill, **G** Surface tamping, **H** Weld complete, **I** Clamp removal. (Color figure online)

that further reduction of cycle time reduced the mechanical properties of welds. This observation from previous authors is observed in the overall data for AA7075-T6 presented in Fig. 3.

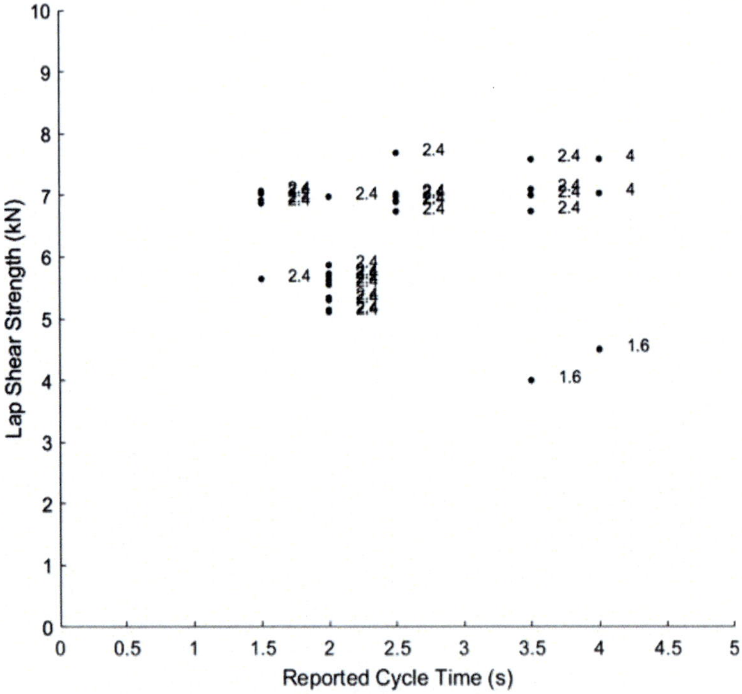

Fig. 3 Plot showing the relationship between RFSSW cycle time and lap-shear strength of RFSSW joints in aluminum alloy 7075-T6, and sheet thickness is displayed in millimeters next to each indicator

Methodology

All welding was completed on a Bond Technologies RFSSW robotic end effector, shown in Fig. 4A, with a H-13 steel tool set with a shoulder diameter of 7 mm (Fig. 4B). Welding was performed on AA7075-T6 with a nominal thickness of 1.6 mm according to the layout shown in Fig. 5. Testing to evaluate the effect of cycle time on weld properties were evaluated via lap-shear tensile tests.

Results and Discussion

To evaluate the ability to further reduce cycle time below previously reported values for RFSSW of AA7075-T6, a nominal set of parameters was used based on previous experience and reported parameters in literature. A rotational velocity of 3000 RPM and a plunge depth of 2.0 mm was selected with a high and low variation for each. Table 1 summarizes the weld parameters evaluated for a fixed cycle time of 1.0 s. The

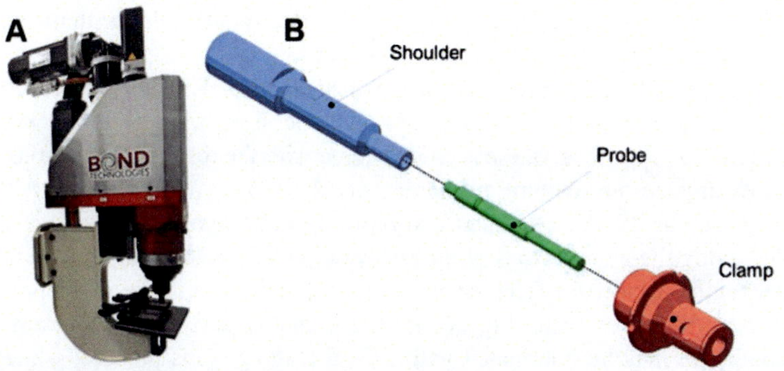

Fig. 4 Depiction of Bond Technologies RFSSW end effector, and associated 7 mm diameter tool set. (Color figure online)

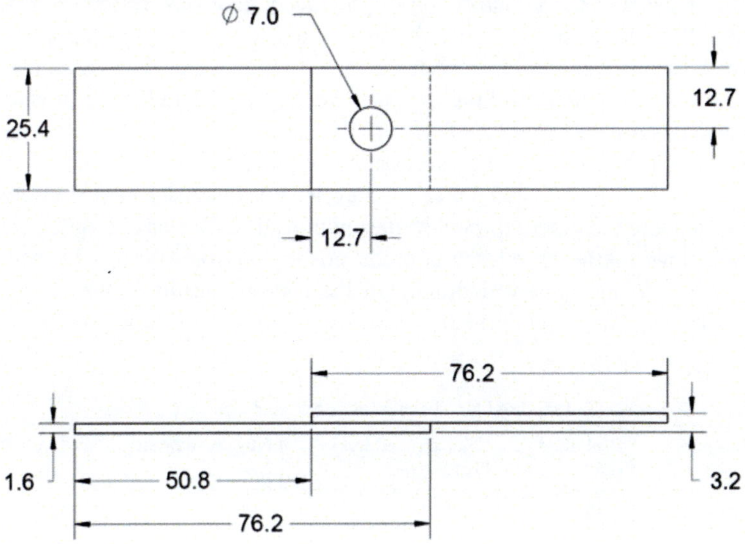

Fig. 5 Specimen geometry and layout for RFSSW lap-shear coupons

Table 1 Weld Parameters for RFSSW of AA7075-T6 with a fixed 1.0-s cycle time

Plunge depth (mm)	Rotational velocity (RPM)	Tamp (Y/N)	Lap-Shear load (kN)	Standard dev (kN)
1.9	3000	Y	5.53	1.04
2	2700	Y	5.74	1.08
2	3000	Y	5.45	0.18
2	3000	N	4.28	0.35
2	3300	Y	5.39	0.47
2.1	3000	Y	5.39	0.71

tamping stage, represented in Fig. 2G, H, was used on all welds with the exception of a repeat of the weld at the nominal conditions.

The most significant influence on lap-shear load of RFSSWs produced with a 1.0 s cycle time as shown in Table 1 was whether a tamping cycle was completed after the weld. This observation seems to agree with the results of numerous authors who investigated the fracture mechanisms and mixing of RFSSW in AA7075-T6. Previous studies [2, 4] concluded that more rapid plunges could lead to a lack of mixing along the vertical wall of the stir zone, which led to a significant reduction in mechanical properties. The purpose of the tamping sequence is to apply a final forging of the mixed weld nugget into the parent material, such that any lack of consolidation may be eliminated with a final forging. This process, while largely unreported in literature, is regularly used in industrial practice to improve surface finish and mechanical performance of the joint.

As most research to date has concluded that further reduction of weld cycle times would be deleterious, a comparison of previously published results is presented in Table 2. Both peak lap-shear strength and normalized lap-shear strength are presented since all results were not done with the same tool diameter. The current work uses a 7 mm tool diameter, resulting in approximately a 40% reduction in welded area from welds produced using a 9 mm tool diameter.

As shown in Table 2, when accounting for variation in weld nugget diameter, welds produced in AA7075-T6 at 1.0 s cycle times actually increased in strength when compared to other reported studies in this alloy. To evaluate further reduction in cycle time, a series of welds were produced ranging from 1000 ms down to 700 ms, see Table 3. While some variation in lap-shear tensile results is shown, there is less than 10% total variation in the lap-shear results for a 30% reduction in cycle time.

Table 2 Comparison of reported lap-shear results and their weld cycle times

Reference	Peak load (kN)	Normalized load (kN/mm^2)	Nugget diameter (mm)	Cycle time (s)
Current	5.45	9.00	7.00	1.00
[4]	7.68	4.60	9.00	2.50
[2]	7.25	4.40	9.00	3.00
[6]	6.00	3.60	9.00	2.00

Table 3 Evaluation showing the influence of reducing cycle time below 1.0 s on lap-shear tensile properties

Cycle time (s)	Plunge depth (mm)	Rotational velocity (RPM)	Tamp (Y/N)	Lap-shear load (kN)	Standard dev. (kN)
1.00	2.00	3000	Y	5.71	0.34
0.90	2.00	3000	Y	5.60	0.46
0.80	2.00	3000	Y	5.17	0.11
0.70	2.00	3000	Y	5.44	0.52

Table 4 Comparison of the influence of reducing cycle time below 1.0 s on tool forces

Cycle time (s)	Probe force (kN)	Standard dev. (kN)	Shoulder force (kN)	Standard dev. (kN)
1.00	6.90	0.21	6.87	0.24
0.90	6.95	0.55	7.14	0.30
0.80	6.72	0.35	7.59	0.17
0.70	7.02	0.38	8.15	0.18

Welds at each cycle time reported in Table 3 demonstrated fracture via nugget pull-out. While the data supports an overall reduction in lap-shear strength for weld cycle times below 1.0 s, the strengths reported still exceed the normalized strengths of those previously reported at 2.0 to 3.0 s.

As noted in several studies, one of the primary concerns related to reducing cycle time is the hardware capability, the machine, and tools, to be able to accommodate for the additional loads, torque, and power required to heat and forge the material quickly. Table 4 compares the machine forces on the probe and shoulder for weld cycle times ranging from 1.0 s to 0.7 s. While the probe forces are essentially the same over this range (less than a 3% variation from the baseline 1.0 s weld) the shoulder forces experience an increase of more than 15% from the baseline. For a 7 mm diameter tool set, buckling or tensile failure of the probe is of greatest concern based on a finite element examination of the tooling. However, at this point the loads are within acceptable limits even at the lowest cycle times.

Conclusions

An investigation focused on demonstrating that cycle times for RFSSW of AA7075-T6 could be reduced below 2.0 s was presented. The data showed that cycle time for RFSSW of AA7075-T6 may be dropped below previously reported cycle times without a commensurate reduction in lap-shear tensile properties. Additional development demonstrated that cycle time could be dropped below 1.0 s, with a reduction by as much as 30% down to 700 ms. A slight reduction in mechanical properties was observed, showing that a 30% reduction in cycle time led to approximately an 8% reduction in peak tensile strength of the welds in AA7075-T6. This additional reduction in cycle time was demonstrated to be machine specific and was largely attainable based on the drive speed and force capability of the Bond RFSSW end effector used for this study. Furthermore, it was observed that tool forces on the probe remained uniform while reducing the cycle time; however, a 15% increase in shoulder forces was observed. Finally, we conclude that while numerous authors have presented that a reduction in cycle time would have an adverse effect on mechanical properties, RFSSWs may be produced at cycle times under 1.0 s without significant reduction in mechanical properties.

References

1. Larsen B, Hovanski Y (2020) Reducing cycle times of refill friction stir spot welding in automotive aluminum alloys. SAE International
2. Shen Z et al (2013) Microstructure and failure mechanisms of refill friction stir spot welded 7075–T6 aluminum alloy joints. Mater Des 44:476–486
3. Shen Z et al (2015) Influence of processing parameters on microstructure and mechanical performance of refill friction stir spot welded 7075–T6 aluminium alloy. Sci Technol Weld Join 20(1):48–57
4. Kubit A et al (2018) Analysis of the mechanical properties and of micrographs of refill friction stir spot welded 7075–T6 aluminium sheets. Archiv Civil Mech Eng 18(1):235–244
5. Kubit A et al (2017) Failure mechanisms of refill friction stir spot welded 7075–T6 aluminium alloy single-lap joints. Int J Adv Manuf Technol 94(9–12):4479–4491
6. Andres J et al (2018) Effect of process parameters on microstructure and mechanical properties of rfssw lap joints of thin AL7075-T6 sheets. Arch Metall Mater 63(1):39–43
7. Boldsaikhan E et al (2017) Refill friction stir spot joining for aerospace aluminum alloys. In: Friction stir welding and processing IX, pp 237–246
8. Boldsaikhan E et al (2016) Refill friction stir spot joining rivet replacement technology. In: SAE Technical Paper Series
9. Liu Z, Yang K, Yan D (2019) Refill friction stir spot welding of dissimilar 6061/7075 aluminum alloy. High Temp Mater Proces (London) 2019(38):69–75
10. Yang, HG, Yang HJ (2013) Experimental investigation on refill friction stir spot welding process of aluminum alloys. In: 2013 3rd international conference on mechanical engineering, industry and manufacturing engineering, MEIME 2013, June 22–June 23, 2013, Trans Tech Publications Ltd., Wuhan, China
11. Xu Z et al (2018) Refill friction stir spot welding of 5083-O aluminum alloy. J Mater Sci Technol 34(5):878–885
12. Yue Y et al (2017) Effect of sleeve plunge depth on microstructure and mechanical properties of refill friction stir spot welding of 2198 aluminum alloy. J Mater Eng Perform 26(10):5064–5071
13. Zhao YQ et al (2014) Effects of sleeve plunge depth on microstructures and mechanical properties of friction spot welded alclad 7B04-T74 aluminum alloy, pp 40–46
14. Kubit A, Wydrzynski D, Trzepiecinski T (2018) Refill friction stir spot welding of 7075–T6 aluminium alloy single-lap joints with polymer sealant interlayer. Compos Struct 201:389–397
15. Wang XJ et al (2014) Comparison of properties on withdrawing and refill friction stir spot welding joints. Appl Mech Mater 633–634:601–606
16. Qiao F et al (2015) An experimental investigation on the dissimilar joining of AA6061 and 1Cr18Ni9Ti by refill friction stir spot welding and its mechanical properties. Proc Institut Mech Eng, Part B: J Eng Manuf 230(4):779–785
17. de Castro CC et al (2018) Taguchi approach for the optimization of refill friction stir spot welding parameters for AA2198-T8 aluminum alloy. Int J Adv Manuf Technol 99(5–8):1927–1936
18. Schmal C, Meschut G, Buhl N (2019) Joining of high strength aluminum alloys by refill friction stir spot welding (III-1854-18). Weld World 63(2):541–550
19. Kwee I, De Waele W, Faes K (2019) Weldability of high-strength aluminium alloy EN AW-7475-T761 sheets for aerospace applications, using refill friction stir spot welding. Weld World 63(4):1001–1011
20. Tier MD et al (2013) The influence of refill FSSW parameters on the microstructure and shear strength of 5042 aluminium welds. J Mater Process Technol 213(6):997–1005
21. Li Z et al (2016) Fracture mechanism of refill friction stir spot-welded 2024–T4 aluminum alloy. Int J Adv Manuf Technol 86(5–8):1925–1932
22. Zhou L et al (2018) Process parameter optimization in refill friction spot welding of 6061 aluminum alloys using response surface methodology. J Mater Eng Perform 27(8):4050–4058
23. Shen Z et al (2019) Microstructure, static and fatigue properties of refill friction stir spot welded 7075–T6 aluminium alloy using a modified tool. Sci Technol Weld Join 24(7):587–600

24. Shen Z et al (2014) Influence of processing parameters on microstructure and mechanical performance of refill friction stir spot welded 7075–T6 aluminium alloy. Sci Technol Weld Join 20(1):48–57
25. Cao JY et al (2016) Hook formation and mechanical properties of friction spot welding in alloy 6061–T6. J Mater Process Technol 230:254–262
26. Cao JY et al (2017) Microstructure, texture and mechanical properties during refill friction stir spot welding of 6061–T6 alloy. Mater Charact 128:54–62

Characterization of Intermetallics Formation in μFSSW of Dissimilar Al/Cu Alloy Sheets

David Yan and Logan Vahlstrom

Abstract Electric vehicle battery packs consist of a large number of battery cells which must be assembled together with robust mechanical and electrical joints. One of the challenges is to join highly conductive and dissimilar multilayer materials such as Al to Cu alloy sheets with 100% reliability. Micro-friction stir spot welding (μFSSW) is a spot-like joining process utilizing a non-consumable tool to generate frictional heat and join multilayer materials under the solid-state condition, which makes the μFSSW a strong candidate for joining dissimilar Al/Cu sheets to manufacture various types of battery interconnects. However, investigations to the dissimilar μFSSW of ultra-thin Al to Cu alloy sheets (<500 μm) are limited. This paper presents experimental investigations on intermetallics (IMCs) formation in μFSSW of ultra-thin Al to Cu sheets. Accordingly, a series of μFSSW of dissimilar 400 μm ultra-thin 6061Al-T6 to Cu-Astm B370 sheets trials were conducted and the stir zone temperature was measured. The obtained weld cross sections were characterized by scanning electron microscopy (SEM) equipped with energy dispersive X-Ray spectroscopy (EDS). The effect of μFSSW conditions on the weld zone IMCs formation was evaluated. It has been found that the IMCs formations are closely related to the processing temperature, which is decided by the tool dwell time in this present study.

Keywords Micro-friction stir spot welding · Dissimilar joining · Microstructure · Intermetallics · Al to Cu alloy sheets

Introduction

In automotive lithium-ion battery manufacturing, a battery pack consists of several hundreds or even more than a thousand battery cells based on the cell configuration to meet the power and energy requirement of the powertrain system in a vehicle. Currently, several joining technologies are utilized in battery manufacturing, including resistance welding, laser welding, ultrasonic welding, and mechanical

D. Yan (✉) · L. Vahlstrom
Department of Aviation and Technology, San José State University, One Washington Sq, San Jose, CA 95192, USA
e-mail: david.yan@sjsu.edu

Y. Hovanski et al. (eds.), *Friction Stir Welding and Processing XI*, The Minerals, Metals & Materials Series, https://doi.org/10.1007/978-3-030-65265-4_19

joining [1]. The joining process of a battery pack is composed of three levels: cell, module, and pack, as illustrated in Fig. 1. The most frequently observed materials in battery interconnecting are aluminum (Al) and copper (Cu) alloys, according to a recent review [2]. The battery tab joint is a critical component of the battery pack and can cause failure of the entire system if it breaks due to impact or fatigue [2]. Presently, this type of weld is usually carried out by ultrasonic welding, which is an excellent technique for dissimilar materials, thin and multilayered sheets, and shows low heat affected zone. However, this technique has disadvantages, including a limitation in joint thickness (<3 mm), sensitivity to product/process variations such as surface conditions, structural vibrations, and severe knurl perforation at the top and bottom of the weld surface [1].

As reviewed, Al and Cu alloys have been widely used in many electronic interconnects due to their good plasticity, high thermal conductivity, and excellent electrical conductivity. Friction stir spot welding (FSSW), as an efficient solid-state welding method suitable for joining of dissimilar metallic materials such as Al to Cu alloys has great prospects in future industrial applications [3, 4]. Li et al. [5] and Zhou et al. [6] performed FSSW of 2 mm thick commercial 1060 Al and T2 Cu plates and found that all the joints possess a Cu hook extruded upward from the lower Cu plate into the upper Al plate with intermetallic compounds (IMCs) developed between the interface. They also found that increasing the dwell time promoted IMCs growth and at short dwell time, the interface was characterized by the interruptedly distributed $CuAl_2$ layer partially mingled with CuAl phase [5]. The IMCs formation sequence for $CuAl_2$, CuAl, and Al_4Cu_9 was predicted according to thermodynamic principles, which is consistent with the interfacial microstructure evolution identified by the transmission electron microscopy [6]. The thickness of IMCs is one of the main factors affecting the weld quality. Momeni and Ni [7] developed a model to predict

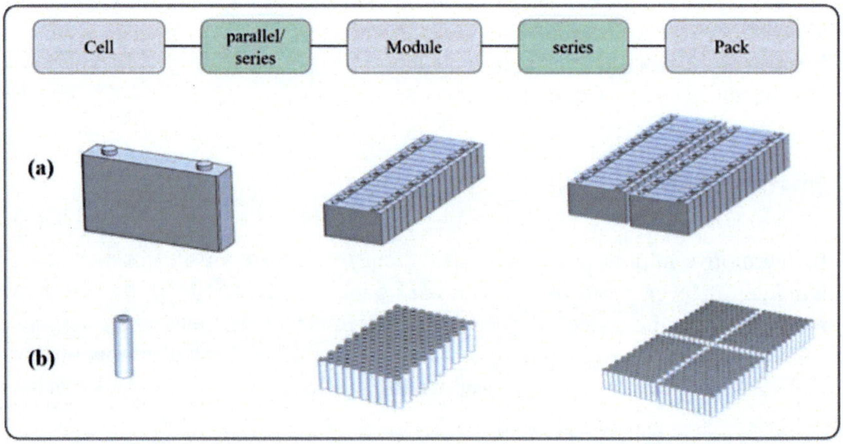

Fig. 1 Overview of battery packs indicating two constructions with (**a**) prismatic and (**b**) cylindrical cells [2]. (Color figure online)

the IMCs thickness in FSW process to improve the understating of the influence of IMCs on the weld integrity.

Up to present, micro-friction sir welding (μFSW) has attracted more attention from scientists and engineers in scientific communities and industrial sectors, and it is another promising solid-state welding technique capable of joining materials with thickness less than 1000 μm [8]. Huang et al. [9] successfully performed μFSW of ultra-thin 6061Al-T4 sheets with thickness of 0.5 mm and proved that the μFSW is feasible and desirable for joining of the ultra-thin Al sheets. For joining battery cells, Mypati et al. [10] joined a 0.3 mm thick Cu sheet to a 0.2 mm thick Al sheet by μFSW, where Cu and Al are used as anode and cathode in adjacent cells for the series arrangement to form a battery pack. They reported that IMCs such as Al_4Cu_9, $AlCu_4$, and AlCu have been found at the joint interface.

As for the micro-friction stir spot welding (μFSSW), it is a spot-like joining process utilizing a tool shoulder, with or without tool pin, to generate frictional heat and join multilayer materials under the solid-state condition. However, μFSSW of ultra-thin dissimilar Al to Cu alloy sheets (<500 μm) is limited and has not been systematically investigated. In this study, experimental studies were performed on μFSSW of ultra-thin 6061Al-T6 and Cu-Astm B370 sheets, both with a thickness of 400 μm. The produced dissimilar Al/Cu weld cross sections were characterized by SEM and EDS to reveal the effect of tool dwell time on the weld zone IMCs formation during μFSSW.

Experimental Procedure

A HAAS CNC mill (TM-1) was employed to perform μFSSW trials with a custom-made rigid clamping and stainless steel backing. A non-tilted tool, including three scrolls flat shoulder with diameter of 15 mm and cylindrical thread pin at diameter of 6 mm and length of 0.48 mm was made of heat-treated H13 tool steel and used in this present study. Commercial 6061Al-T6 (composition in weight %: Mg 1.17, Si 0.57, Cu 0.26, Fe 0.45, Cr 0.22, Al remaining) and Cu-Astm B370 sheets (99.9% Cu), both with the thickness of 400 μm and width of 25 mm were selected to form dissimilar Al-to-Cu workpieces used in μFSSW trials. These two sheets were overlapped by 25 mm in length with the Al sheet on the top and μFSSW was conducted in the center of the overlapped area. A series of μFSSW trials was conducted when the tool was rotating at a constant rotational speed (ω) of 1000 rpm and plunging speed of 4 mm/min (V) with tool dwell time (T) ranging from 10 to 20 s. A constant 0.22 mm depth of shoulder plunging (D) was used in all experiments. Figure 2 shows the experimental setup, a scroll tool and a μFSSWed sample embedded beneath the fixture.

The stir zone temperature was measured by embedding a tip of 0.127 mm diameter thermocouple wire (OMEGA 5TC-TT-E-36–72) at the faying surface of the Al and Cu sheets, with 4 mm away from the welding centerline. After each μFSSW trial, the produced Al-to-Cu weldments were sectioned in the transverse direction through

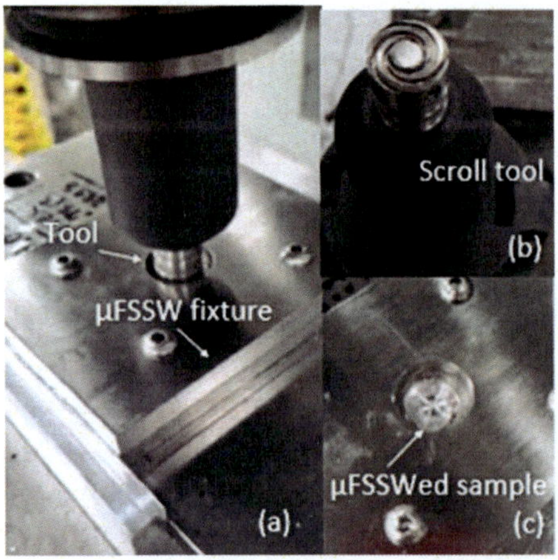

Fig. 2 Photos of (**a**) μFSSW experimental setup, (**b**) scroll tool, and (**c**) μFSSWed sample embedded beneath the fixture. (Color figure online)

the welding centerline. All cross-sectional samples were mounted and polished with standard metallographic practice followed with 0.3 μm Alumina–water slurry for final polishing, and etched using Kroll's etchant (2 ml HF and 6 ml HNO_3 in 100 ml H2O). Subsequently, the etched samples were examined via optical microscope and SEM equipped with EDS.

Results and Discussion

Post-Weld Profile and Cross-Sectional Configuration in μFSSW

Figure 3(a), (b), (c), (d) shows two post-weld profiles and lower magnification SEM micrographs of weld cross sections made using scroll tool with dwell time for 10 and 15 s, respectively, at a constant rotational speed of 1000 rpm, plunging speed of 4 mm/min and 0.22 mm depth of shoulder plunging. Observation of these two post-weld profiles shows that surface-defect-free welds were achieved under the current μFSSW conditions. Examination using SEM in Fig. 3(b), (d) has shown the common features in μFSSW such as stir zone (SZ) and hook structures. It can be found that a Cu hook extruded upward from the lower Cu sheet into the upper Al sheet, and the shape of the hook varied due to different dwell time as depicted in the figure. In fact, at shorter dwell times, a smaller hook formed at the interface due to the insufficient heat input compared to a larger hook developed at a longer dwell time.

It can also be seen from Fig. 3(b), (d) that there was a void that appeared in the weld cross section obtained at the tool dwell time of 10 s but there was a void-free weld made when increasing the tool dwell time from 10 to 15 s. The trend of the effect of tool dwell time on the Cu hook formation and the weld cross-sectional

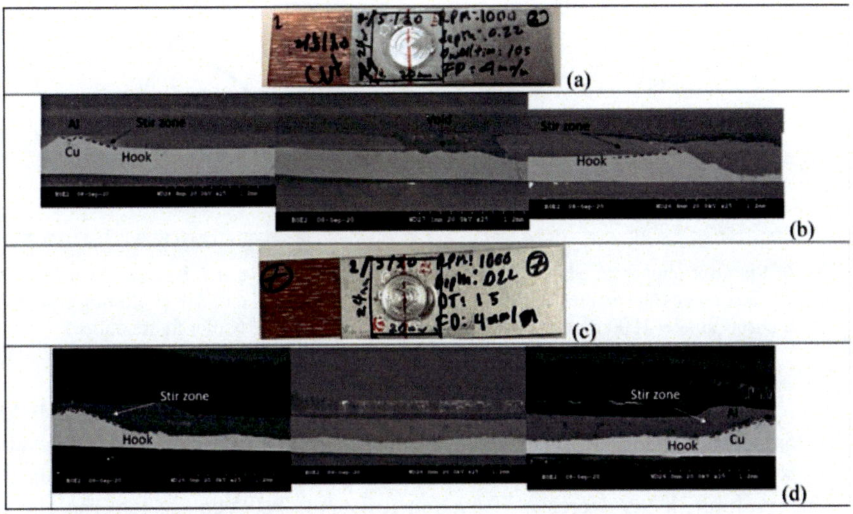

Fig. 3 Photos of two post-weld profiles obtained at (**a**) T = 10 s and (**c**) T = 15 s, and two SEM micrographs of weld cross sections produced at (**b**) T = 10 s, showing void occurred within the stir zone, and (**d**) T = 15 s, showing void-free weld during μFSSW of ultra-thin dissimilar Al/Cu sheets at ω = 1000 rpm, V = 4 mm/min and D = 0.22 mm. (Color figure online)

features were similar to the findings of Li et al. [5], although the greater thickness of 2 mm 1060 Al and T2 Cu sheets were used in their study. Apparently, a continuous metallurgical bonding could be achieved in the form of IMCs layer along the Al to Cu interface.

Stir Zone Temperature History in μFSSW

It is necessary to investigate the thermal history during μFSSW as the elevated temperature is the most important μFSSW parameter for softening workpiece, mixing stir zone materials and nucleation, and growth of IMCs, for which their evolutions are thermally activated [5, 6]. The above two welds' stir zone temperature histories corresponding to the tool dwelling time of 10 and 15 s were plotted in Fig. 4(a), (b). It can be seen from Fig. 4(b) that the peak temperature of 300 °C was recorded at tool dwell time of 15 s, while around 260 °C of peak temperature as shown in Fig. 4(a) was reached at the tool dwelling of 10 s. Thus, an increase in tool dwell time leads to a higher peak temperature that can be concluded from this present study.

The stir zone temperature curves shown in Fig. 4(a), (b) have a similar variation trend, i.e. the increasing stage is characterized by two-steep rises. This is because when the bottom surface of the scroll tool pin contacted the upper surface of the Al sheet, the drastic friction occurred between these two surfaces and thus a large

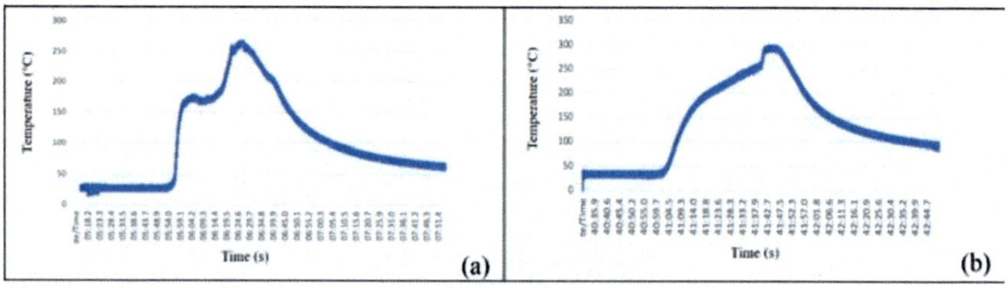

Fig. 4 Stir zone temperature history during μFSSW at (**a**) T = 10 s and (**b**) T = 15 s showing the effect of tool dwell time on the peak temperature of stir zone during μFSSW of ultra-thin dissimilar Al/Cu sheets at ω = 1000 rpm, V = 4 mm/min and D = 0.22 mm. (Color figure online)

amount of friction heat is generated, which could be responsible for the first steep of temperature rise. However, when the temperature exceeded a threshold value, the workpiece materials were softened and began to flow around the rotating tool pin, resulting in the decrease of friction resistance between the tool pin and the workpiece materials, thus, the rate of the temperature rise decreased. In terms of the second steep temperature rise, as the scroll tool was continuously plunging into the upper Al sheet, the tool shoulder contacted the Al sheet, which increased the frictional area significantly and led to the second steep of the temperature rise.

When comparing the stir zone peak temperature generated at tool dwelling time of 10 and 15 s to the weld cross-sectional morphologies as shown in Fig. 3(b), (d), it appeared that the stir zone temperature affected the weld zone materials/layers development that could consequently lead to the formation of a void or void-free weld. Thus, the higher the weld zone temperature, the better quality of the produced μFSSW weld. The funding of this present study was consistent with the temperature measurement reported by Zhou et al. [6].

Stir Zone Al/Cu IMCs Formation During μFSSW

For comparison, Fig. 5(a), (c) presents high magnification SEM micrographs of two void-free stir zones as shown in Fig. 3(b), (d), while Fig. 5(b), (d) is higher magnification SEM micrographs of area A marked in Fig. 5(a), (c), respectively. It can be seen from Fig. 5(a), (c) that no crack or substantial void is present in these two welds, although there is a small local discontinuity along the interface region.

Examination using higher magnification SEM has also shown that there is a little localized IMCs micro-crack in the weld produced at tool dwell time of 10 s as shown in Fig. 5(b), but there is no substantial crack from the weld obtained at the tool dwell time of 15 s as shown in Fig. 5(d). It appears that the tool dwell time corresponding to the stir zone temperature affected the IMCs crack formation. Furthermore, by examining Fig. 5(b), there are two layers formed within the IMCs section, the outermost layer, i.e. marked as layer 2 can extensively grow outward toward Al matrix.

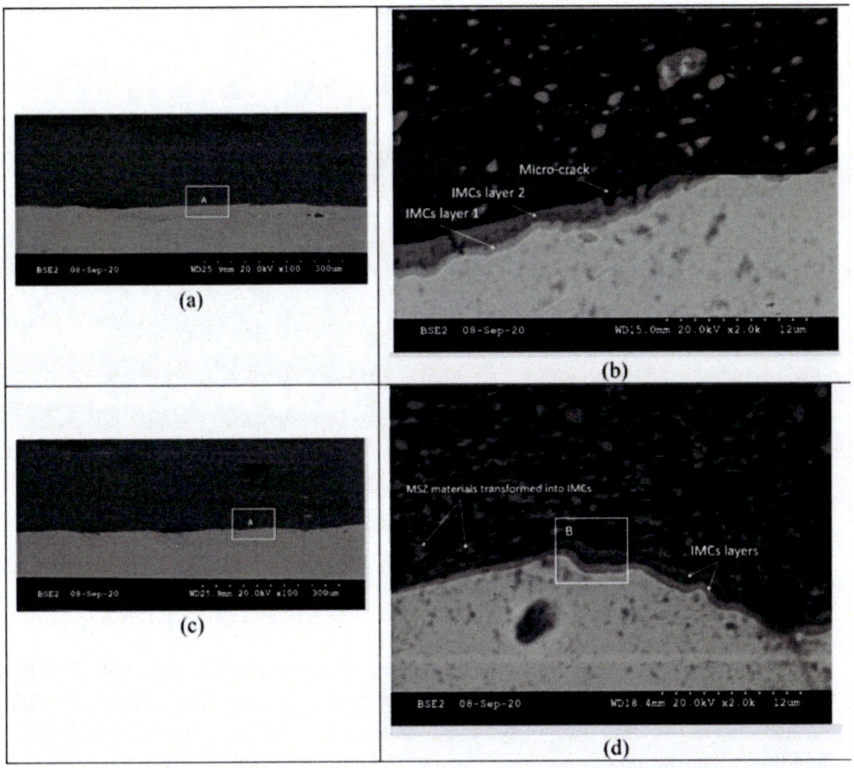

Fig. 5 (**a**) and (**c**) high magnification SEM micrographs of two void-free stir zones shown in Fig. 3(b), (d), showing Al to Cu bonding interface, and (**b**) and **d** higher magnification SEM micrographs of area A marked in Fig. 5(a), (c), respectively, showing IMCs layers formed along the Al to Cu interface

This indicates that the Al down flow materials can be driven downward by the tool pin and forced to weld with IMCs to establish a metallurgical joint. Moreover, by examining Fig. 5(d), it can be deduced that some mix stir zone (MSZ) materials have transformed into IMCs due to excessive frictional heat as indicated in the figure.

Growth Behavior of Al/Cu IMCs During μFSSW

Figure 6(a) shows a higher magnification SEM micrograph of area B marked in Figs. 5(d) and 6(b) which present EDS spectra for the selected areas A and B marked in Fig. 6(a), respectively. By measuring the IMCs thickness from Fig. 6(b), (d), it appears that the formations of the IMCs are closely related to the temperature, which is decided by the tool dwell time in this present study. At the shorter tool dwell time of 10 seconds, the interface was characterized by a 1.9 μm thick laminated IMCs

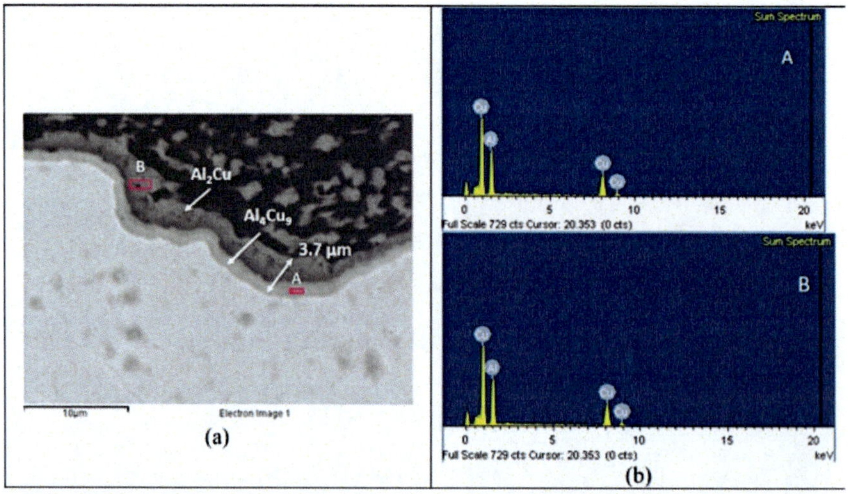

Fig. 6 (**a**) higher magnification SEM micrograph of area B marked in Fig. 5(d) and (**b**) EDS spectra for the selected areas A and B marked in pink in Fig. 6(a), showing chemical compositions of these two IMCs layers. (Color figure online)

layer with two sub-layers shown in Fig. 6(b). While at the longer dwell time of 15 s, the IMCs thickness was 3.7 μm measured from Fig. 6(a). Accordingly, it appears that the rise in the thickness of the IMCs layer at the Al/Cu interface was a result of the higher heat input.

In terms of the chemical compositions of these two IMCs layers, i.e. areas A and B marked in Fig. 6(a), the results of EDS analysis shown in Fig. 6(b) indicated that the compositions are approximately equal to that of the Al_4Cu_9 (area A) and Al_2Cu (area B), measured from the selected areas A and B. This result was consistent with the finding of Li et al. [5].

Conclusions

The ultra-thin dissimilar Al/Cu sheets (400 μm thick) were successfully joined utilizing μFSSW. The produced weld cross sections were characterized by SEM and EDS to reveal the effect of tool dwell time on the weld zone IMCs formation during μFSSW. Based on the results obtained from this present study, the following conclusions can be drawn:

1. Two μFSSW welds all possess a Cu hook extruded upward by the tool pin from the lower Cu sheet into the upper Al sheet with IMCs developed on Al/Cu interface.
2. At shorter dwell times, a smaller hook is formed at the interface due to the insufficient heat input compared to a larger hook developed at longer dwell times.

3. The thickness of the IMCs layer varies between 1.9 and 3.7 μm.
4. The IMCs' chemical compositions are approximately equal to that of the Al_2Cu and Al_4Cu_9.

Acknowledgements The authors acknowledge the financial support of the Undergraduate Research Grants of San Jose State University.

References

1. Lee S et al (2010) Joining technologies for automotive lithium-ion battery manufacturing. Review 1
2. Zwicker MFR et al (2020) Automotive battery pack manufacturing – a review of battery to tab joining. J Adv Joining Process 1:100017
3. Li M et al (2020) Friction stir spot welding of aluminum and copper. Rev Mater 13(1):156
4. Mubiayi MP, Akinlabi ET, Makhatha ME (2018) Current state of friction stir spot welding between aluminium and copper. Mater Today: Proc 5(9, Part 3): 18633–18640
5. Li G et al (2019) Influence of dwell time on microstructure evolution and mechanical properties of dissimilar friction stir spot welded aluminum–copper metals. J Mater Res Technol 8(3):2613–2624
6. Zhou L et al (2019) Microstructure evolution and mechanical properties of friction stir spot welded dissimilar aluminum-copper joint. J Alloy Compd 775:372–382
7. Momeni F, Ni J (2020) A new model for predicting the thickness of intermetallic compounds in friction stir welding. J Manuf Process
8. Khan HA, Wang K, Li J (2018) Interfacial bonding mechanism and mechanical properties of micro friction stir blind riveting for multiple Cu/Al ultra-thin layers. Mater Charact 141:32–40
9. Huang Y et al (2019) Microstructures and mechanical properties of micro friction stir welding (μFSW) of 6061–T4 aluminum alloy. J Mater Res Technol 8(1):1084–1091
10. Mypati O et al (2020) A study on electrical and electrochemical characteristics of friction stir welded lithium-ion battery tabs for electric vehicles. J Electron Mater 49(1):72–87

Temperature Distribution During Friction Stir Spot Welding of Thin AA 6082-T6 and AA 5082-O Sheets

Mikhail A. Ozhegov, Fedor Yu. Isupov, and Roman I. Smelianskii

Abstract Friction Stir Spot Welding (FSSW) is a solid-state joining technique which is widely used in the automotive, aircraft, and other industries. The heat treatable aluminum alloy—6082-T6—and strain hardenable aluminum alloy—5082-O—were used as a material for the experimental procedure. In the present research, the optimization of the FSSW parameters was performed to obtain the joints with the highest mechanical characteristics. After an optimization procedure, three sets of parameters for FSSW of each alloy were evaluated to obtain process temperature using thermocouples.

Keywords Friction stir spot welding · Temperature distribution · Thermocouples · Aluminum alloys · Thin sheets

Introduction

Automotive and aerospace industries have begun integrating FSSW into production lines and facilities. Several automotive companies, such as Mazda Motor Corporation, Toyota Motor Corporation, and Honda Motor Corporation, use FSSW at their manufacturing line to join different materials [1, 2]. FSSW allows them to implement aluminum into the car bodies and to reduce the weight of the bodies and increase the production efficiency with the required quality. By means of FSSW not only similar materials could be joined, but also dissimilar material sets [3–5]. For instance, aluminum alloys can be welded with steel and carbon fiber composite and be successfully used in automotive components.

The microstructure and mechanical properties of FSSWed joints were presented and analyzed by the authors [6–8]. Only a few articles describe the thermal cycles and temperature ranges in FSSW [9, 10]. The authors of [11–14] presented the

M. A. Ozhegov · F. Yu. Isupov (✉) · R. I. Smelianskii (✉)
Peter the Great St. Petersburg Polytechnic University, St. Petersburg, Russia
e-mail: isupov_fyu@spbstu.ru

R. I. Smelianskii
e-mail: smel9n@gmail.com

Y. Hovanski et al. (eds.), *Friction Stir Welding and Processing XI*, The Minerals, Metals & Materials Series, https://doi.org/10.1007/978-3-030-65265-4_20

temperature cycles during FSW for AA 6082-T6. the authors of [15–17] analyze the temperature influence, mechanical properties, and material behavior during the friction stir welding process of the aluminum alloy 5082-O. All this research shows a great potential to continue the work in this field of FSSW. Based on the wide usage of aluminum alloys 6082-T6 and 5082-O in the automotive industry, the thermal cycle measurement process during FSSW was realized and presented.

Experimental Procedure

In the present research, the aluminum alloys 6082-T6 and 5082-O are used. The materials of the 5XXX and 6XXX series are characterized by high fracture toughness, good corrosion resistance, and good cold formability. These alloys are mainly used in car body construction, aerospace, and rail vehicle construction. The chemical composition, physical and mechanical properties of these alloys are presented in Tables 1, 2, 3, and 4.

Two square sheets with dimensions of 10 × 10 cm were used as samples. Previously, the samples were processed with a degreaser. The samples' geometry and welding configurations can be seen in Fig. 1a.

Four thermocouples were welded to the workpiece on a pulse-arc welding machine Orion 150S on the upper workpiece and secured with aluminum tape to reduce

Table 1 Chemical composition of the 6082-T6 aluminum alloy

Element	Al	Mg	Fe	Zn	Si	Mn	Cr	Cu	Ti
wt. %	95.2–98.3	0.6–1.2	0–0.5	0–0.2	0.7–1.3	0.4–1.0	0–0.25	0–0.1	0–0.1

Table 2 Physical and mechanical properties of the 6082-T6 aluminum alloy

Physical and mechanical properties	6082-T6
Density (g/cm3)	2.7
Melting temperature (°C)	580–650
Tensile strength Rm (N/mm2)	330
Yield strength Rp0.2 (N/mm2)	270
Elongation at break A (%)	9.8
Thermal conductivity (W/mK)	160
Hardness (HB)	93

Table 3 Chemical composition of the 5082-O aluminum alloy

Element	Al	Mg	Fe	Zn	Si	Mn	Cr	Cu	Ti
wt. %	93.5–96	4.0–5.0	0–0.35	0–0.25	0–0.2	0–0.15	0–0.15	0–0.15	0–0.1

Table 4 Physical and mechanical properties of the 5082-O aluminum alloy

Physical and mechanical properties	5082-O
Density (g/cm3)	2.7
Melting temperature (°C)	560–640
Tensile strength Rm (N/mm2)	400
Yield strength Rp0.2 (N/mm2)	340
Elongation at break A (%)	1.1
Thermal conductivity (W/mK)	130
Hardness (HB)	65

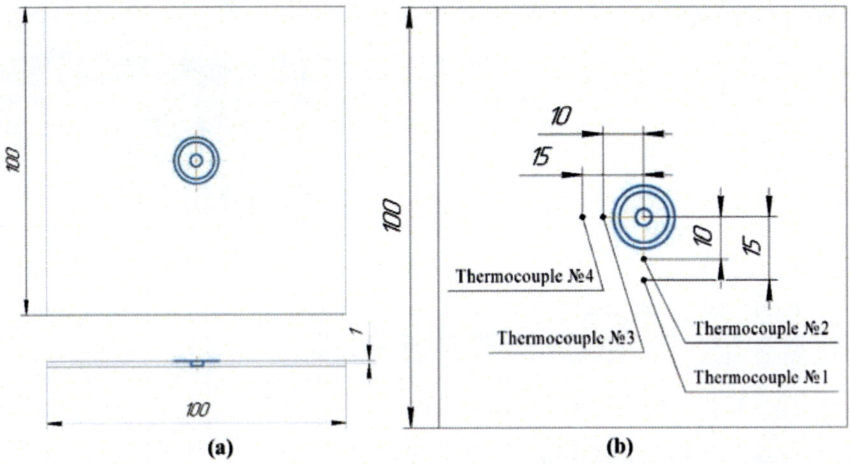

Fig. 1 Samples' geometry **a** and the thermocouples positions **b** (unit: mm). (Color figure online)

their mobility and prevent them from tearing off from the workpiece. To measure thermal cycles, K-type thermocouples with a wire thickness of 0.2 mm were used. The positions of thermocouples are shown in Fig. 1b.

As can be seen in Fig. 1b, the thermocouples are equally distant from the center of weld and are located on two sides for better accuracy.

The welding process was realized on a specialized machine MATEC-40 P. This gantry-type machine is designed for both single and serial production of large and heavy parts for tool production, machine-building production, and metal construction. The main field of application is the three-dimensional processing of steel and aluminum parts, cast and welded steel structures.

Six pairs of samples were prepared for this experiment. Four thermocouples were pre-welded to the six samples. FSSW regimes were developed for six pairs of workpieces (Table 5).

Increasing the number of workpieces and additional modes will allow us to obtain a larger number of temperature cycles for a wider understanding of thermal processes in the material under various welding conditions.

Table 5 FSSW regimes

Weld IDs	1, 2, 3, 4, 5, 6
Rotation speed (rpm)	8000, 4000, 1000
Axial force (kN)	4
Plunge speed (mm/min)	150, 85, 50
Plunge depth (mm)	1.8
Welding time (s)	10
Probe length (mm)	2
Shoulder type	Concave

Fig. 2 NI cDAQ-9184. (Color figure online)

To measure and record the temperature cycles, a four-slot chassis Ethernet Compact DAQ-cDAQ-9184 was used (Fig. 2).

Results and Discussion

Figure 3 shows the temperature distribution during the FSSW of weld ID1 for 6082-T6 aluminum alloy (Table 6), but during preparation a thermocouple fell off the

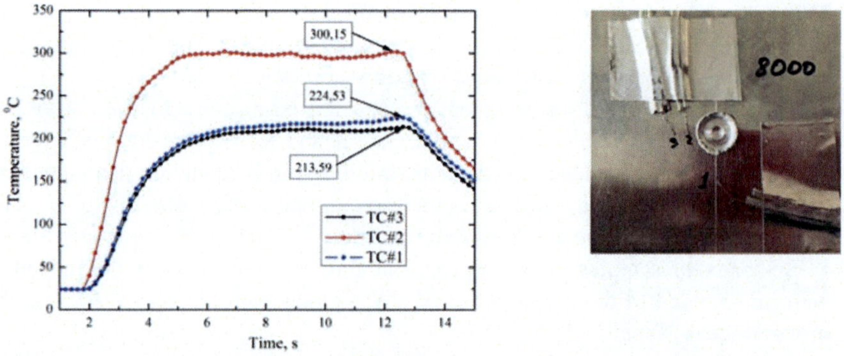

Fig. 3 Temperature distribution during the FSSW of weld ID1. (Color figure online)

Table 6 FSSW of weld ID1

Rotation speed (rpm)	Axial force (kN)	Plunge speed (mm/min)	Plunge depth (mm)	Welding time (s)	Probe length (mm)	Shoulder type
8000	4	150	1.8	10	2	Concave

workpiece, therefore, the graph shows the temperature distribution only for three thermocouples.

Figure 4 shows the temperature distribution during the second welding mode for 6082-T6 aluminum alloy (Table 7).

Figure 5 shows the temperature distribution during the third welding mode for 6082-T6 aluminum alloy (Table 8); during the welding process, one thermocouple did not show any results and one more fell off the workpiece, so the graph shows the temperature distribution only in two out of four thermocouples.

Figure 6 shows the temperature distribution during the FSSW of fourth welding mode for 5082-O aluminum alloy (Table 9).

Figure 7 shows the temperature distribution during the FSSW of the fifth welding mode for 5082-O aluminum alloy (Table 10); the graph shows the temperature distribution for four thermocouples.

Figure 8 shows the temperature distribution during the FSSW of the sixth welding mode for 5082-O aluminum alloy (Table 11), but during the preparation for welding

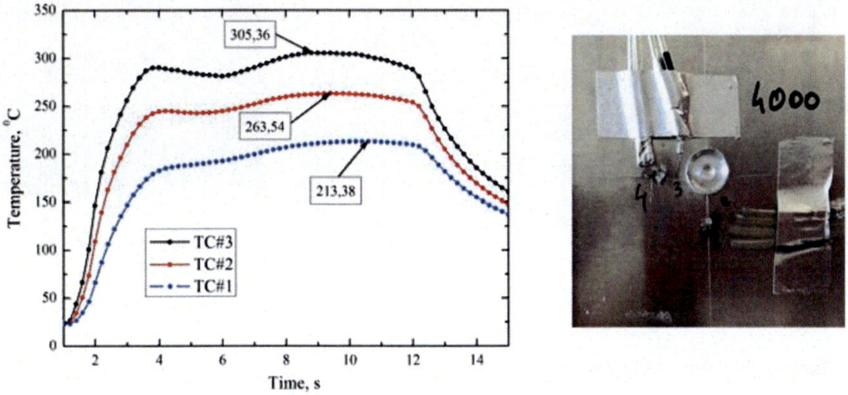

Fig. 4 Temperature distribution during the FSSW of weld ID2. (Color figure online)

Table 7 FSSW of weld ID2

Rotation speed (rpm)	Axial force (kN)	Plunge speed (mm/min)	Plunge depth (mm)	Welding time (s)	Probe length (mm)	Shoulder type
4000	4	85	1.8	10	2	Concave

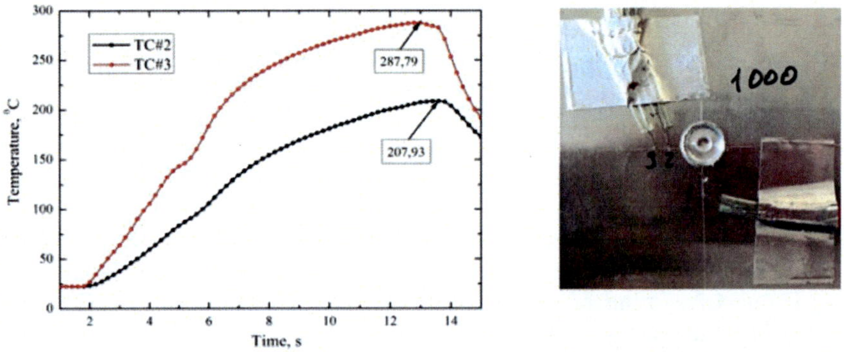

Fig. 5 Temperature distribution during the FSSW of weld ID3. (Color figure online)

Table 8 FSSW of weld ID3

Rotation speed (rpm)	Axial force (kN)	Plunge speed (mm/min)	Plunge depth (mm)	Welding time (s)	Probe length (mm)	Shoulder type
1000	4	50	1.8	10	2	Concave

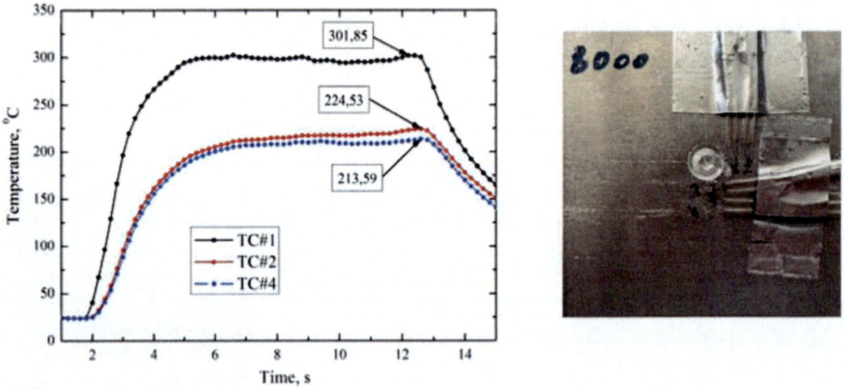

Fig. 6 Temperature distribution during the FSSW of weld ID4. (Color figure online)

Table 9 FSSW of weld ID4

Rotation speed (rpm)	Axial force (kN)	Plunge speed (mm/min)	Plunge depth (mm)	Welding time (s)	Probe length (mm)	Shoulder type
8000	4	150	1.8	10	2	Concave

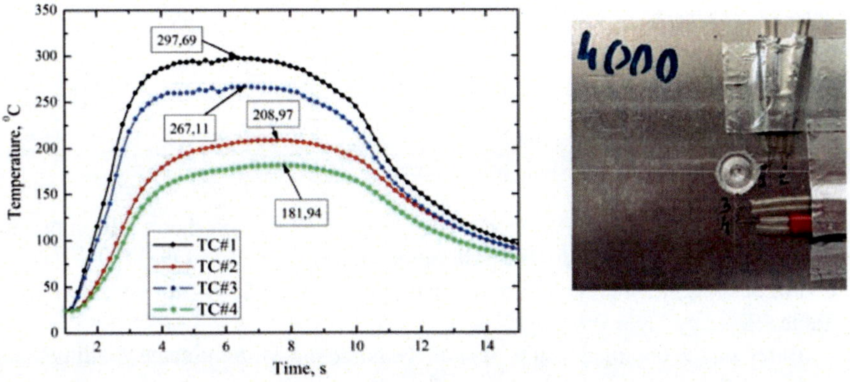

Fig. 7 Temperature distribution during the FSSW of weld ID5. (Color figure online)

Table 10 FSSW of weld ID5

Rotation speed (rpm)	Axial force (kN)	Plunge speed (mm/min)	Plunge depth (mm)	Welding time (s)	Probe length (mm)	Shoulder type
4000	4	85	1.8	10	2	Concave

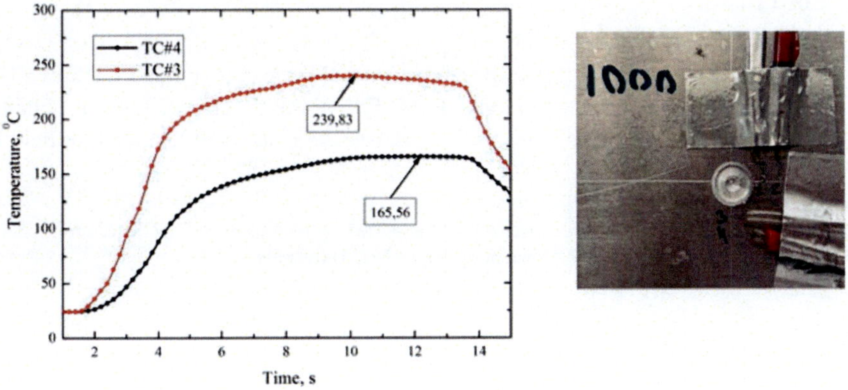

Fig. 8 Temperature distribution during the FSSW of weld ID6. (Color figure online)

Table 11 FSSW of weld ID6

Rotation speed (rpm)	Axial force (kN)	Plunge speed (mm/min)	Plunge depth (mm)	Welding time (s)	Probe length (mm)	Shoulder type
1000	4	50	1.8	10	2	Concave

Table 12 Maximum temperatures, °C

AA 6082-T6		AA 5082-O	
ID 1	300.15	ID 4	301.85
ID 2	305.36	ID 5	297.69
ID 3	287.79	ID 6	239.83

two thermocouples fell off the workpiece, therefore, the graph shows the temperature distribution only in two out of four thermocouples.

For each mode, the maximum temperatures for the FSSW process are shown in Table 12.

Based on the obtained results, it could be noted that with the same welding modes (Tables 6–11) and the same distance of the thermocouples from the weld center, in the aluminum alloys discussed above, a similar behavior of temperature distribution is observed (Figs. 3, 4, 5, 6, 7 and 8). The maximum temperatures measured during FSSW are slightly higher for the AA 6082-T6. It should be also noted that with an increasing of rotation speed, the maximum temperatures increase significantly (Figs. 6, 7 and 8).

Conclusion

In this paper, the process of measuring the thermal cycles for aluminum alloys 6082-T6 and 5082-O by six welding modes during FSSW was described. The data obtained in the present research will be used to validate the calculated temperature field by finite element modeling.

Acknowledgements This research work was supported by the Academic Excellence Project 5-100 proposed by Peter the Great St. Petersburg Polytechnic University.

References

1. Iwashita T (2003) Method and apparatus for joining, US6601751B2
2. Kusuda Y (2013) Honda develops robotized FSW technology to weld steel and aluminum and applied it to a mass production vehicle. Ind Robot: Int J 40(3): 208 – 212
3. Haghshenas M, Gerlich AP (2018) Joining of automotive sheet materials by friction-based welding methods: a review. Eng Sci Technol Int J 21:130–148
4. André NM, dos Santos JF, Amancio-Filho ST (2019) Evaluation of joint formation and mechanical performance of the AA7075-T6/CFRP spot joints produced by frictional heat. Materials 12(891): 1–14
5. Boeing 787: From the Ground Up. Available online: www.boeing.com/commercial/aeromagazine/articles/qtr_4_06/article_04_2.html. Accessed 12 May 2017
6. Shen Z, Yang X, Yang S, Zhang Z, Yin Y (2014) Microstructure and mechanical properties of friction spot welded 6061–T4 aluminum alloy. Mater Des 54:766–778

7. Shen Z, Yang X, Zhang Z, Cui L, Li T (2014) Microstructure and failure mechanisms of refill friction stir spot welded 7075–T6 aluminum alloy joints. Mater Des 44:476–486
8. Su, J-Q, Nelson TW, Mishra R, Mahoney M (2003) Microstructural investigation of friction stir welded 7050-T651 aluminum. Acta Mater 51:713–729
9. Chen K, Liu X, Ni J (2019) A review of friction stir–based processes for joining dissimilar materials. Int J Adv Manuf Technol 104:1709–1731
10. Hatzky M, Frank A, Böhm S (2019) Friction stir spot welding with additional bonding of thick sheet aluminum joints. Metals 9(732):1–11
11. Naumov A, Morozova I, Rylkov E, Obrosov A, Isupov F, Michailov V, Rudskoy A (2019) Metallurgical and mechanical characterization of high-speed friction stir welded AA 6082–T6 aluminum alloy. Materials. 12:4211
12. Golubev I, Morozova I, Naumov A, Hantelmann C, Doynov N, Michailov V (2017) Numerical simulation and experimental investigation on impulse friction stir welding of 6082-t6 aluminum alloy. In: Contributed Papers from Materials Science and Technology (MS&T17), pp 987–994
13. Golubev IA, Chernikov EV, Naumov AA, Michailov VG (2016) Temperature distribution and welding distortion measurements after FSW of Al 6082-T6 sheets. In: Friction stir welding and processing VIII, pp 289–295
14. Chernikov EV, Naumov AA, Golubev IA, Michailov VG (2015) Numerical simulation and microstructure analysis of friction stir welded Al 6082. In: METAL 2015—24th international conference on metallurgy and materials, conference proceedings, pp 1423–1428
15. Naumov A, Isupov F, Golubev I, Morozova I, Spree C (2018) Development of methodology to determine the temperature influence on microstructure evolution during friction stir welding. In: Contributed Papers from Materials Science and Technology 2018 (MS&T18), pp 795–802
16. Naumov A, Chernikov E, Isupov F, Panchenko O (2017) Friction stir welding of dissimilar in thickness Al-5Mg alloy butt joints. In: Contributed Papers from Materials Science and Technology (MS&T17), pp 971–977
17. Isupov F, Naumov A, Panchenko O, Zhabrev L, Popovich A (2018) Comparative analysis of the mechanical properties of aluminum alloys welded joints obtained by friction stir welding. In: Contributed Papers from Materials Science and Technology (MS&T18), pp 867–874

Author Index

Y. Hovanski et al. (eds.), *Friction Stir Welding and Processing XI*, The Minerals, Metals & Materials Series, https://doi.org/10.1007/978-3-030-65265-4

Subject Index

Y. Hovanski et al. (eds.), *Friction Stir Welding and Processing XI*, The Minerals, Metals & Materials Series, https://doi.org/10.1007/978-3-030-65265-4